ITCA Information Technology Certified Associate

DATA SCIENCE FUNDAMENTALS

Study Guide

Part of the ITCA Stackable Certificate Program
www.isaca.org/itca

ISACA®

About ISACA

For more than 50 years, ISACA® (*www.isaca.org*) has advanced the best talent, expertise and learning in technology. ISACA equips individuals with knowledge, credentials, education and community to progress their careers and transform their organizations, and enables enterprises to train and build quality teams. Among those credentials, ISACA advances and validates business-critical skills and knowledge through the globally respected Certified Information Systems Auditor® (CISA®), Certified in Risk and Information Systems Control® (CRISC®), Certified Information Security Manager® (CISM®), Certified in the Governance of Enterprise IT® (CGEIT®) and Certified Data Privacy Solutions Engineer™ (CDPSE™) credentials. ISACA is a global professional association and learning organization that leverages the expertise of its 145,000 members who work in information security, governance, assurance, risk and privacy to drive innovation through technology. It has a presence in 188 countries, including more than 220 chapters worldwide.

Disclaimer

ISACA has designed and created the *Data Science Fundamentals Study Guide* (the "Work") primarily as an educational resource for professionals. ISACA makes no claim that use of any of the Work will assure a successful outcome. The Work should not be considered inclusive of all proper information, procedures and tests or exclusive of other information, procedures and tests that are reasonably directed to obtaining the same results. In determining the propriety of any specific information, procedure or test, professionals should apply their own professional judgment to the specific circumstances presented by the particular systems or information technology environment.

Reservation of Rights

ISACA
1700 E. Golf Road, Suite 400
Schaumburg, IL 60173, USA
Phone: +1.847.660.5505
Fax: +1.847.253.1755
Contact us: *https://support.isaca.org*
Website: *www.isaca.org*

Provide Feedback: To comment, please visit *https://support.isaca.org*.
Participate in the ISACA Online Forums: *https://engage.isaca.org/onlineforums*

Twitter: *http://twitter.com/ISACANews*
LinkedIn: *www.linkedin.com/company/isaca*
Facebook: *www.facebook.com/ISACAGlobal*
Instagram: *www.instagram.com/isacanews/*

ISBN 978-1-60420-755-2
Data Science Fundamentals Study Guide
Printed in the United States of America

Acknowledgments

The *Data Science Fundamentals Study Guide* is the result of the collective efforts of many volunteers. Special thanks go to:

Lead Developer

Sarit Williams, COBIT, CSM, CSTP, PMP, 360Adept.com, USA

Expert Reviewers

Hafiz Sheikh Adnan Ahmed, CGEIT, Certified COBIT 5 Assessor, Certified DPO, UAE
Adham Etoom, CRISC, CISM, FAIR, GCIH, PMP, Jordan
Jason Otani, CISM, USA
Malgorzata Pankowska, Poland
Xitij Urmil Shukla, Ph.D., CISA, India
Melroy Patrick Vanlangenberg, Australia
Guodong Zou, CISA, CRISC, CISM, CGEIT, CBAP, EBDP, Six Sigma Black Belt, China

Page intentionally left blank

Table of Contents

Page intentionally left blank

LIST OF FIGURES

Chapter 1. Data Science Concepts and Overview

Chapter 2. Data Management

Chapter 3. Data Science Process

Introduction

The *Data Science Fundamentals Study Guide* is a reference guide designed to assist individuals in acquiring a base-level knowledge of data science in preparing to earn the ISACA® Data Science Fundamentals certificate, one of the components of the ISACA Information Technology Certified Associate™ certification program.

Data science, regarded as a newer branch of science, is a cross-pollination of statistics, computer science and business disciplines. Data science addresses the needs of businesses to exploit their mass data collection for better sales, optimized operations and improved customer service, thus increasing organizational value and easing consumer frustration with mundane tasks.

The main goal is to enable the inventory of assets digitally via data collection. Data science employs artificial intelligence and machine learning to analyze raw data for relationships and to bring context to it. For example, applications such as LinkedIn or Facebook use data science tools to suggest new connections or friends automatically, based on users' existing contacts and interactions.

Mass data collection poses new challenges to understanding what value data may hold; whom it can serve; where it can be used and stored, and for what purpose; and how it can be ethically managed for all involved. This guide will explore these questions and offer guidance, including highlights relevant to each of the component disciplines, and to the integration of the three to create what is known as data science today.

Topics in this guide are presented at a foundational level and include basic terminology, concepts, examples, general practices and explanations of networks and their infrastructure. The *Data Science Fundamentals Study Guide* provides the IT or business professional with knowledge in the following main content areas:

- Data science concepts and overview
- Data management
- Data science process

Every chapter contains the following:

- **Learning objectives**—Knowledge that learners should understand after reading the chapter
- **Sections**—Knowledge content
- **Knowledge check**—Quiz on the chapter knowledge

Page intentionally left blank

DATA SCIENCE
FUNDAMENTALS
An ISACA Certificate

Chapter 1:

Data Science Concepts and Overview

Data Science Concepts and Overview

Data Science Concepts and Overview

1.1 Learning Objectives

After completing this chapter, learners will be able to:

1. Define the terms and concepts of data science.
2. Describe the classifications and characteristics of data.
3. Explain the different types of data structures, flows and diagrams.
4. Describe the relationship between data science and statistics.
5. Distinguish among sampling techniques.
6. Use statistical analysis to gather populations and samples.
7. Explain the role of machine learning in data science.

1.2 Data Characteristics

1.2.1 What Is Data Science?

Data science is a new branch of science used to extract knowledge and insights from large and complex data sets. The data science domain frequently requires knowledge of statistics, software engineering and business analysis, with added depth in each depending on the role within the data science discipline. With the advent of electronic data collection, and its use and storage for general services, it became clear that stored data could be extracted and analyzed to form the basis of new, more focused marketing and other services that would allow enterprises to make better decisions for consumers and stakeholders alike.

1.2.2 Defining Big Data

Big data refers to very large, complex data sets with data variety that increases in volume at higher rates of velocity.

This definition is based on an understanding of the original three Vs: variety, volume and velocity. This concept has been expanded to encompass the five Vs of big data:

- **Volume**—This "V" is the scale or size of the data or the amount of data generated.
- **Velocity**—This "V" refers to the speed of data generation and the rate at which it is processed. Velocity is the analysis of moving or streaming data.
- **Variety**—This "V" refers to the many types of data, which come from many input sources.
- **Veracity**—This "V" refers to the quality of the data. Veracity is defined by the degree of uncertainty or accuracy of data. (Is the data accurate and reliable?)
- **Value**—This "V" is concerned with how effectively the enterprise's management will leverage data to make decisions. Value refers to the benefits of understanding data and applying it to achieving business goals.

Volume, velocity, variety, veracity and value are the five keys to understanding big data.

1.2.3 The Evolution of Big Data

The main concepts of big data are always evolving. It is likely that the current understanding will expand beyond the five Vs as big data is redefined in the future (see **figure 1.1**). Some possible additions to the Vs are the following:

- **Validity**—This "V" refers more specifically to the precision and accuracy of data. This is where most data scientists spend time. Typically, data scientists spend 60 percent of their time cleaning up data.[1]
- **Vulnerability**—This "V" relates to cybersecurity risk. The main concern is how data might be compromised.
- **Volatility**—This "V" refers to how quickly data becomes irrelevant and invalid. This is typically one of the hardest questions to answer regarding Big Data. Data retention is often poorly understood. Retaining data for a period of time can be costly and may be subject to a myriad of privacy-related regulatory considerations.
- **Visualization**—This "V" represents the many ways to view big data. Charts and graphs provide a visual understanding of large data sets, allowing for succinct conveyance of information to nontechnical audiences.

Figure 1.1—Overview of Big Data's Vs

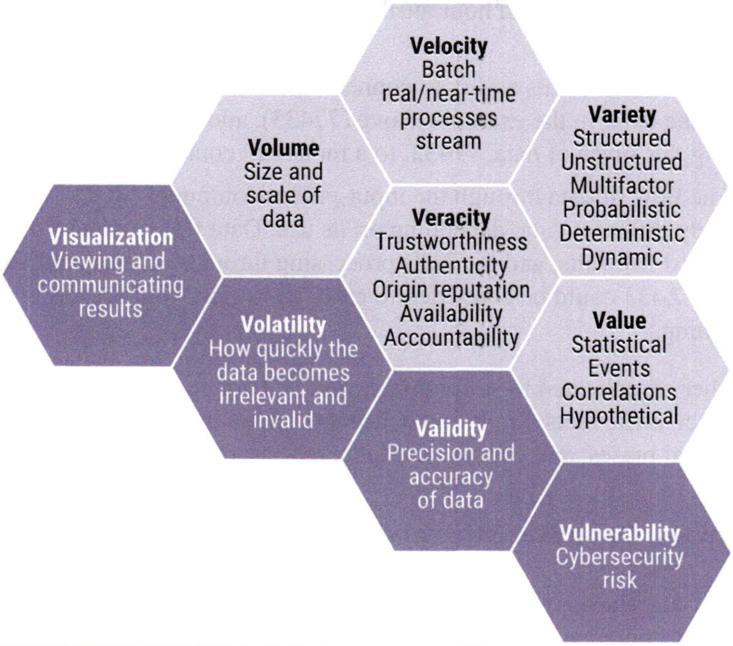

Source: Adapted from Moura, J.; C. Serrão; "Security and Privacy Issues of Big Data," *Handbook of Research on Trends and Future Directions in Big Data and Web Intelligence*, 1st edition, ed. S.N. Zaman, F. Seliaman, M.F. Hassan, F. Garcia Marquez, IGI Global, 2 September 2015, figure 1, pg. 23, *https://www.igi-global.com/book/handbook-research-trends-future-directions/124160*

1.2.4 What Is Data?

Data, in the simplest terms, is a piece of information that has no value other than itself and is self-referencing with no external indications. It stands alone. As a term, data is also defined as facts, according to some dictionaries. However, as this guide will explain, data that is manipulated or simply contextualized can be proven to lack integrity. Data integrity, which is a top enterprise priority, is the expectation that the data quality is reliable. For example, the number 77433 on its own can mean a slew of things: a ZIP code, a sum total of aggregate numbers, or the number for a home address.

Data may be cataloged as either quantitative or qualitative. Quantitative, related to quantity, indicates how much or how many of something. Qualitative, related to quality, denotes data categories, names, labels or associations.

[1] Press, G.; "Cleaning Big Data: Most Time-Consuming, Least Enjoyable Data Science Task, Survey Says," *Forbes*, 23 March 2016, *www.forbes.com/sites/gilpress/2016/03/23/data-preparation-most-time-consuming-least-enjoyable-data-science-task-survey-says/?sh=742fbda16f63*

1.2.5 Raw Data vs. Contextualized Data

As a piece of independent information, data is raw. It has no context—in other words, it has no meaning or insight. Data science's pure value is in creating meaning, that is, drawing conclusions from independent data by making sense of it. For data to have meaning it must have dimensionality, connectivity and transformability.

- **Dimensionality**—In statistics, dimensionality refers to how many attributes a dataset has. The time data dimension may include days, hours, minutes, seconds, etc. A dimension reduction technique is used to extract one or more dimensions that capture as much of the variation in the data as possible. Using the time example, the reduction may be to include date and hour only, removing all other time dimensions. Dimension reduction allows for a reduction of the volume and complexity of the data mass while keeping the essence of what the data is attempting to convey (e.g., the date and hour attributes will suffice in this case while the seconds attribute may provide too much information).
- **Connectivity**—The independent data's ability to connect to other markers, concepts or variables may change the context, or meaning, of the data. In the example above (77433), adding the currency symbol ($) will result in $77,433, thus attributing this piece of data, 77433, to a monetary context.
- **Transformability**—The flow of data through the input-process-output cycle will cause the data to transform into similar or completely different data, depending on its use. Data flow includes travel through communication lines, routers, switches and firewalls, and it entails processing through various applications, websites, banks and the like. For example, $77,433 could be converted to euro (€), which would result in €68,249.446, providing a different value, or meaning.

Contextualized data is further complicated by additional layers that are not as defined and straightforward as mathematical symbols and scientific facts. Signs and concepts that are indisputable by most may be affected by layers of perception, individual biases, and cultural and industry-based contexts that are more fluid and less defined.

1.2.6 Difference Between Data Statistics and Analytics

Prior to the evolution of the data science field, data analytics and data statistics were two very distinct areas of study and application. Statistics, a mathematical branch of science, is concerned with collecting, analyzing, presenting and interpreting data. Government needs for census data and information about a variety of economic activities provided much of the early impetus for the field of statistics. Generally, statisticians collect small samples of data and use rigorous statistical formulas to search for patterns and trends. They ask questions and perform high-level analyses.

The two disciplines have become more difficult to distinguish, as their tasks and end goals are quite blended. Data science, however, is often understood as a broader, more task-driven and computation-oriented version of statistics.[2]

Additional distinct differences are detailed below:

- **Prediction vs. inference**—Prediction is the ability to review existing data and develop estimates as to future data results or behaviors (e.g., artificial intelligence) while inference draws conclusions and stops there. Inference does not influence future outcomes or decision-making directly, as it is more theoretical. A statistician is concerned with theory and understanding how data works, while a data scientist is concerned with creating something that does the work—*solves a problem.*
- **Exploratory data analysis vs. confirmatory analysis**—Data science is mostly exploratory data analysis. Within statistics, exploratory data analysis (EDA) contrasts with confirmatory analysis (i.e., searching for hypotheses vs. attempting to confirm a hypothesis). EDA can become problematic when an analyst puts too much faith in its results (i.e., mistakes hypothesis generation for hypothesis confirmation). Every statistician can list problems with simple correlation data set mining (e.g., false discovery, sampling bias). Identification of such

[2] Carmichael, I.; Marron, J.S.; "Data science vs. statistics: two cultures?," *Japanese Journal of Statistics and Data Science*, vol. 1, 2018, 117-138, *https://doi.org/10.1007/s42081-018-0009-3*

problems does not mean the results are wrong, but it indicates that exploratory analysis provides much weaker evidence for a hypothesis than confirmatory analysis. How much this matters is context dependent.[3]

- **Problem first vs. problem search**—Data science focuses on problem solving—that is, making data analysis useful to other disciplines.[4] Statistics, on the other hand, focuses on the development of procedures and the study of results to identify problems relevant to the findings.[5]

Statisticians use data analysis and data scientists use statistics. Most competent data scientists have a good grasp of statistical tools, and some statisticians have experience with programming languages. This guide covers both disciplines as they apply to the data science body of knowledge.

1.2.7 Data Types

Any discussion of data must first delineate fundamental differences among the types of data one may encounter in the data science field. Data types are important because statistical methods and programmatic manipulation can be used only with certain data types. For example, arithmetically adding string values (i.e., alphabetical characters) will result in a NaN (not a number), as strings are textual.

Figure 1.2 is a broad overview of the data types referenced throughout this guide.

Figure 1.2—Overview of Data Types

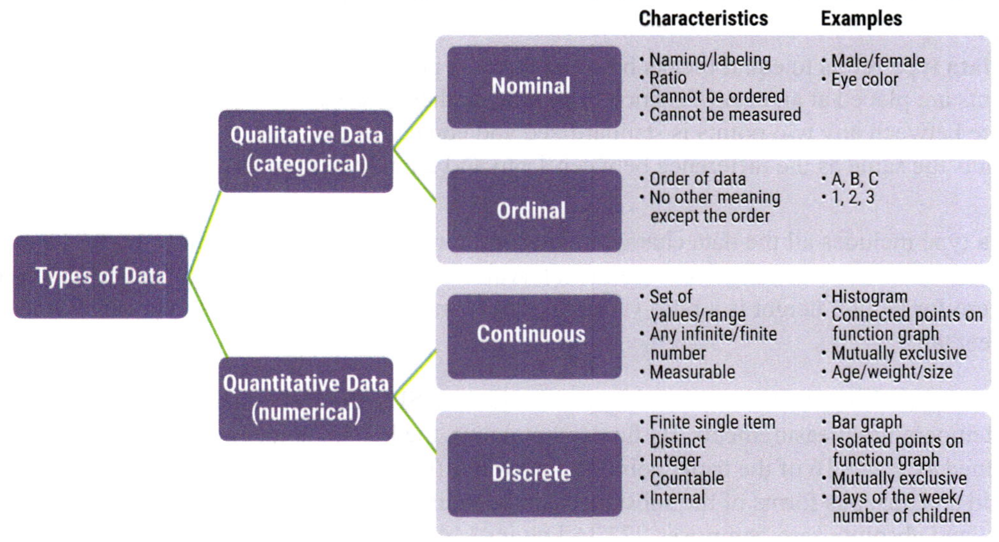

The performance of continuous data analysis must differ from the performance of categorical data analysis. Otherwise, the results would be wrong. Knowing the types of data involved and defining the problem to be solved enables analysts to choose the correct method.

In the 1940s, Harvard psychologist Stanley Smith Stevens created a taxonomy of measurement that identifies four classes of scales—nominal, ordinal, interval and ratio—and distinguishes among them on the basis of the operations that leave the form of the scale invariant.[6] In other words, data is measured on scales that are unchanging and static.

[3] *Ibid.*
[4] *Ibid.*
[5] *Ibid.*
[6] Elsevier Science Ltd., "Steven's Biography," *International Encyclopedia of the Social & Behavioral Sciences, http://web.mit.edu/epl/StevensBiography.pdf*

Stevens' typology has been used in the social sciences, mathematics and statistics, to name a few fields. Data scientists have adopted it partially, especially for use in machine learning with specific data types. Machine learning is a program or system that builds (trains) a predictive model from input data. The system uses the learned model to make useful predictions from new (never-before-seen) data drawn from the same distribution as the one used to train the model. Machine learning also refers to the field of study concerned with these programs or systems.[7]

Following is a description of each of the data types.

- **Nominal data** type refers to discrete values that have no numerical meaning and no numerical relationships among the different categories. Mean and median are meaningless. Surveys often code such data with numbers, e.g., 0 = male and 1 = female. In some cases, nominal data may be numbers, but the numbers have no numerical meaning. For example, truck is not higher than van or lower than jeep. Other examples include a person's gender (M, F) or possible colors of a new car.
- **Ordinal data** type refers to discrete data-like integers that can be ranked or sorted. Integers represent a range of mathematical numbers that are whole numbers and are commonly represented in a computer as a group of binary digits (bits). Although ordinal data can include whole numbers, the numbers themselves, or the distances between them, have no meaning. Meaning is derived from ordering, ranking or sorting. For example: How likely are you to recommend this computer to a friend? 5 = absolutely, 4 = certainly, 3 = maybe, 2 = rarely, 1 = not at all.
- **Continuous data** type is not limited to integers and can include a floating-point data type, e.g., height. A floating-point is a number that has fractional parts and is usually expressed with a decimal point, e.g., someone's weight may be 126.3 lbs.
- **Interval data** type refers to equal breaks between the data points. Interval data are measured along a scale, in which points are placed at an equal distance from one another. Interval data always appear in numerical form; the distance between any two points is standardized and equal. For example, the difference between 5 and 6 is one, which is the same as the difference between 1 pm and 2 pm (one number**).** Ratios are not used with interval data.
- **Ratio data** type includes all the data characteristics of interval data, plus an absolute zero. The number zero means there is none of something. Because the intervals between numbers are comparable and there is an absolute zero for height, height is a ratio variable. Furthermore, a 6-foot-tall person is twice as tall as a person who is 3 feet tall.

  Another example: Using the International System of Units (ISU), the Kelvin (K) scale is the standard for all scientific temperature measurements. Its fundamental unit, the kelvin, is identical in size to the Celsius degree and is defined as 1/273.16 of the triple point of pure water (0.01°C [32.02°F])—i.e., the temperature at which the liquid, solid and gaseous forms of the substance can be maintained simultaneously. The interval between this triple point and absolute zero comprises 273.16 kelvins.[8] In the Kelvin scale for temperatures there is no possible value less than zero degrees Kelvin, so it is an absolute zero measurement (see **figure 1.3**). Kelvin is considered ratio data as negative values are not possible. But temperatures measured in Fahrenheit and Celsius are not ratio data, as negative values are valid.

[7] Google, "Machine Learning Glossary," *https://developers.google.com/machine-learning/glossary/*
[8] Encyclopædia Britannica, "Absolute Zero," 18 January 2019, *www.britannica.com/science/absolute-zero*

Figure 1.3—Ratio Data Example

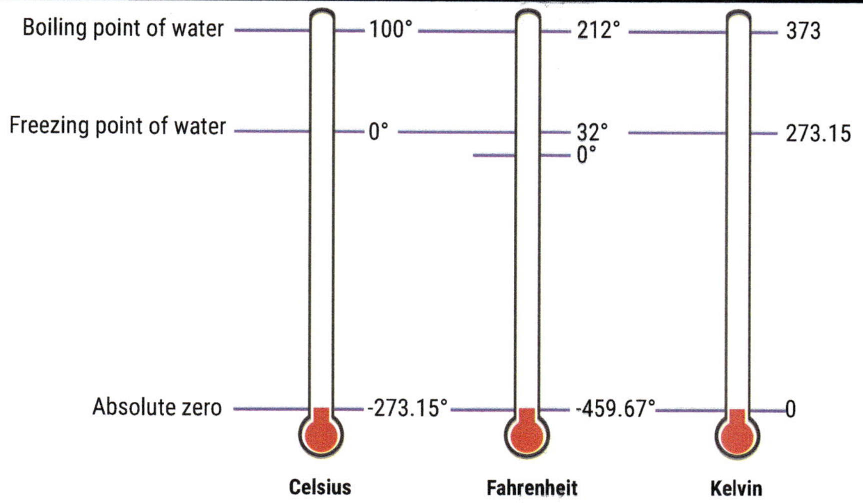

- **Character data** type, also known as "char" in software development, refers to a single alphabetical character string such as A or X (or any other languages). Char can also include special character symbols, such as asterisk (*) or exclamation mark (!) and the any of the Unicode and ASCII spectrum of numbers in the range of 256-65535—only one character at a time.
- **String data** type can be used to identify and hold an indefinite number of characters. An example would be "How are you today?" or "Peace." It includes the spaces as part of the data string and considers it as one value.

1.2.8 ASCII and Unicode

A computer understands only numeric values. Prior to 1963, computer manufacturers had more than 60 different ways of representing characters in computers. Machines could not communicate with one another. This problem became increasingly evident as companies began networking computers.

The International Organization for Standardization (ISO®), the world's largest developer of voluntary international standards, released ISO/IEC 2022 specifying the structure of the 8-bit codes and 7-bit codes that enable the coding of character sets.[9] ISO/IEC 2022 created a standard all could follow to enable networking and provide the foundation for encoding a common programming language.

ASCII characters are used in programming, data conversions, text files, graphic art and emails. Programmers can use ASCII to represent calculations on characters.

ASCII (pronounced "AS-KEE") stands for American Standard Code for Information Interchange. It uses numbers to represent text. Digits (1,2,3, etc.), letters (a, b, c, etc.) and symbols (!) are called "characters." When there is a piece of text, ASCII converts each character to a number (see **figure 1.4**). This set of numbers is easier to store in computer memory. In simple words, assigning a number to a character is called "encoding."

[9] International Organization for Standardization (ISO®), ISO/IEC 2022:1994 *Information Technology – Character Code Structure and Extension Techniques*, December 1994, *www.iso.org/standard/22747.html*

Figure 1.4—ASCII Conversion Table

Character	Binary Code	Character	Binary Code	Character	Binary Code	Character	Binary Code
A	01000001	U	01010101	o	01101111	)	00101001
B	01000010	V	01010110	p	01110000	*	00101010
C	01000011	W	01010111	q	01110001	+	00101011
D	01000100	X	01011000	r	01110010	,	00101100
E	01000101	Y	01011001	s	01110011	-	00101101
F	01000110	Z	01011010	t	01110100	.	00101110
G	01000111	a	01100001	u	01110101	/	00101111
H	01001000	b	01100010	v	01110110	0	00110000
I	01001001	c	01100011	w	01110111	1	00110001
J	01001010	d	01100100	x	01111000	2	00110010
K	01001011	e	01100101	y	01111001	3	00110011
L	01001100	f	01100110	z	01111010	4	00110100
M	01001101	g	01100111	!	00100001	5	00110101
N	01001110	h	01101000	"	00100010	6	00110110
O	01001111	i	01101001	#	00100011	7	00110111
P	01010000	j	01101010	$	00100100	8	00111000
Q	01010001	k	01101011	%	00100101	9	00111001
R	01010010	l	01101100	&	00100110	?	00111111
S	01010011	m	01101101	'	00100111	@	01000000
T	01010100	n	01101110	(	00101000	_	01011111

Unicode is a universal coding system designed for use with different language scripts. There are three types of encoding available in Unicode. They are UTF-8, UTF-6 and UTF-32. UTF-8 uses 8 bits per character, UTF-16 uses 16 bits per character and UTF-32 uses 32 bits per character. In UTF-8, the first 128 characters are the ASCII characters. Therefore, ASCII is valid in UTF-8. Usually, Unicode is used in internationalization and localization of computer software. This standard is also used in operating systems, XML, .NET framework and programming languages such as Java.

ASCII can represent only 128 characters; it includes lowercase letters (a-z), uppercase letters (A-Z), digits (0-9), and symbols such as punctuation marks. Unicode represents a wider range of characters than ASCII, including English, Arabic and Greek letters; mathematical symbols; historical scripts; and emoji.

1.2.9 DIKW Pyramid

The data, information, knowledge and wisdom (DIKW) pyramid, also known as a hierarchical framework, explains the building blocks and subcomponents currently used in most information systems, sciences and management domains. Each discipline knowingly or unknowingly uses this hierarchy along a repeatable continuum to answer several questions about data: what, why and how.

Figure 1.5 displays the DIKW pyramid.

Figure 1.5—Data, Information, Knowledge and Wisdom Pyramid (DIKW)

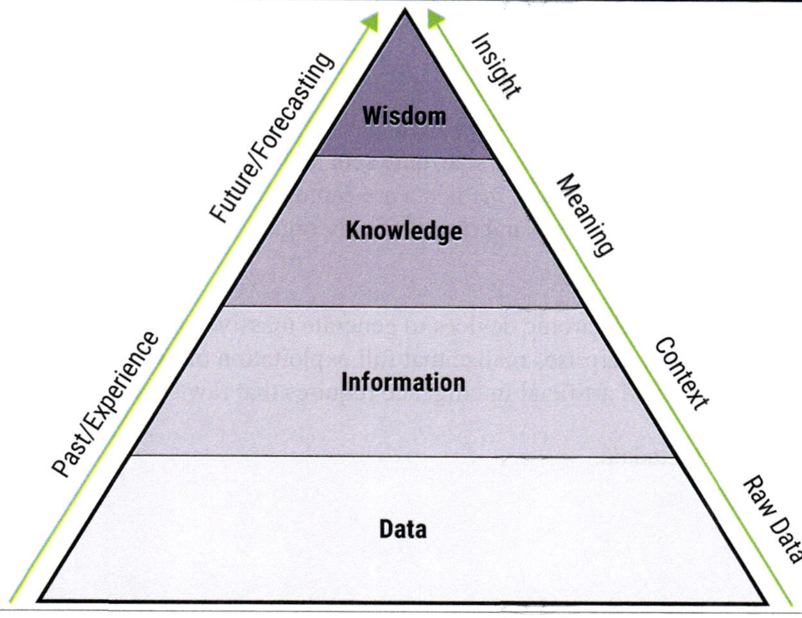

According to the Journal of the American Society for Information Science and Technology, the three layers—data, information and knowledge—are interrelated. Those relationships, and the inferences drawn from them, can and do change conclusions and influence actions.

- **Data**—The data portion of the pyramid refers to raw, unadulterated facts, symbols or signs representing stimuli or signals, either in a preprocessed form and collected according to some algorithm for some purpose, or in a post-processed form, e.g., tables of such information. Postprocessed data are dependent on information and cannot be defined apart from it.[10] It is expected that data have been quantified or qualified according to type for further use.
  - **Garbage in, garbage out (GIGO)** is the concept that nonsensical or flawed data produces nonsensical or flawed output, especially in the computation sciences.
- **Information**—Information refers to the context or relationships identified or expanded when examining raw data. The word "information" is used to refer to a number of different phenomena, classified into three groupings: 1) anything perceived as potentially signifying something (e.g., printed books); 2) anything involved in the process of informing; and 3) anything learned from evidence or communication.[11] As the second level of the hierarchy, information heavily relies on the person or system (i.e., who or what) being informed or how the information is disseminated, which adds another level of complexity when considering final conclusions.
  - Data as symbolism or stimuli are highly dependent on social, cultural and personal biases when combined with information, or a contextual layer. For example, once information is added to 56.3, it may look like this: 56.3 lbs. Further, 56.3 lbs. can be heavy or light depending on context. An adult who weighs 56.3 lbs. may be light; contrarily, a baby who weighs 56.3 lbs. may be heavy. Moreover, 56.3 lbs. means nothing to someone who is versed only in the metric system, which measures weight in kilograms.
- **Knowledge**—Knowledge is actionable information that is further processed into knowingness by comparing and contrasting it, assuming it was received as intended. It is that which is known—by definition subjective, even when aggregated to the level of social, or public, knowledge—the sum, in a sense, of individual "knowings."[12]

[10] Zins, C.; "Conceptual Approaches for Defining Data, Information and Knowledge," *Journal of the American Society for Information Science and Technology,* 15 February 2007, *www.success.co.il/is/zins_definitions_dik.pdf*

[11] *Ibid.*

[12] *Ibid.*

- **Wisdom**—The apex of the pyramid is the ultimate transformation of any rational process. Transforming experience, past thoughts, behaviors, actions or data into meaningful, actionable and valuable next steps in pursuit of a worthy goal beneficial to a person or enterprise—that is wisdom.

1.2.10 Metadata

Metadata are data about data. Metadata describes data attributes via labels, bookmarks or any other tagging method. For example, an email's metadata may include file size, date sent and received, time sent and received, sender and recipient email addresses, and much more. Some metadata are captured programmatically and some data are tagged post-processing when the data are qualitatively and quantitatively organized (e.g., based on folder rules set up by the user for categorization).

In this age of billions of people using electronic devices to generate massive amounts of data, the volume of metadata is growing exponentially. Enterprises realize that full exploitation of the business and operational potential of machine learning, deep learning and artificial intelligence requires that raw data be enhanced with metadata.

There are three main aspects to metadata:

- Observation
- Identifiers
- Tagging

Observation

Observation is needed to collect metadata, whether it is performed programmatically or manually. Observation can occur naturally or organically, or it can be staged to test a hypothesis. Observations can be categorized into qualitative and quantitative data types. Naturally occurring observations rely on collected data. Time and patience are required to collect just the right type of experiential data. Observation is performed via receipt of messages through electronic, sensory or vibrational signals, and the human senses.

Staged observation involves faux data independently created for observation. Software test engineers who are tasked with carrying out the quality assurance test-and-integration phase of the software development life cycle (SDLC) are well acquainted with this skill, because they must create a majority of the test data to ensure the right environment variables are present for observing how the program will behave during specific use cases (e.g., business scenarios).

Identifiers

An identifier is a distinguishing characteristic that makes a piece of data unique. A UID (unique identifier, also known as unique ID) is an attribute (usually a number, code or piece of information, e.g., social security number) that is guaranteed to be unique among all the metadata used for a given piece of data.

In data science, unique identifiers are required to track a piece of information through its processing life cycle, but more so, they are required to make decisions that apply to a grouping of unique identifiers (e.g., all driver licenses in the state of Texas begin with the number two for the UID record in a database).

UIDs require special handling because they can expose sensitive personal information that is well regulated and protected, such as race, age, health stats and religion. Considering sensitive personal information in conclusions can create observational bias.

Data scientists and their employers can avoid observational bias, or at a minimum greatly reduce it and meet regulatory standards, by applying data anonymization programmatically to prevent the dissemination of unique identifiers. This approach protects personal information and individual privacy, while ensuring that data are still valid for the proposed analysis. For example, the finance department may need to know a person's social security, number of dependents, address, etc., to process payroll, but the shift supervisor may need only the name and department ID to process shift schedules. Information can be hidden from view, masked, or shared partially on an as-needed basis.

Based on experiments conducted using 1990 US Census data, it was found that 87 percent (216 million of 248 million) of the population had reported characteristics that likely made them uniquely identifiable based only on 5-digit ZIP, gender, and date of birth.[13]

Employing a data classification level of sensitivity to data (or information) can result in the specification of controls for each level of classification. Levels of sensitivity of data are assigned according to predefined categories as data are created, amended, enhanced, stored or transmitted. The classification level is an indication of the value or importance of the data to the enterprise based on factors such as criticality, sensitivity and ownership.

There are three ways of implementing data anonymity:

- **Encryption**—Encryption is the process of taking an unencrypted message (plaintext), applying a mathematical function to it (encryption algorithm with a key), and producing an encrypted message (ciphertext). Only authorized parties with the key can access it. Those who are not authorized cannot, because they would require the decryption key. A threat can occur in the case of a non-authorized party accessing the encrypted information and decrypting it after obtaining the key.
- **Masking/obfuscation**—Masking is obscuring, hiding or partially replacing data with other data—e.g., instead of showing a full social security number showing only the last four digits (XXX-XX-7888). Obfuscation removes, shuffles or otherwise changes the appearance of data, without the loss of meaning or relationships to other data sets. Masking is also known as pseudonymization.
- **Deletion of source data**—Although it is not available in many projects, when deletion of source data is an option, one can opt to not collect personal identifiable information, or to delete it after a set time period. For example, some double-blind health research projects do not include personal names. Including only relevant attributes such as gender and age is sufficient in a specific context. Another example is requiring only nonsensitive consumer data—such as lifestyle, preferences, search history or age range—when creating profiles for marketing campaigns.

Tags

In information systems, a tag is a keyword or term assigned to a piece of information. It is not the data itself, but rather a descriptor of the data that allows it to be categorized or searched quickly. This kind of metadata helps describe an item and allows it to be found again by browsing or searching. Tags are generally chosen informally and personally by the item's creator or by its viewer, depending on the system, although they may also be chosen from a controlled vocabulary.[14]

Tags and labels for unknown or untagged data are assigned by individuals familiar with the data context, the source, or business roles that may be able to use that data. The employees of the organization make judgments about a given piece of unlabeled data, which is known as curation or label engineering. Once the data is properly classified, machine learning models can be applied to the dataset.

An International Data Corporation (IDC) study projected that from 2013 to 2020, the digital universe would grow by a factor of 10—from 4.4 trillion gigabytes to 44 trillion. It more than doubles every two years.

[13] Sweeney, L.; "Simple Demographics Often Identify People Uniquely," Carnegie Mellon University, Data Privacy Working Paper 3, USA, 2000
[14] Smith, G.; *Tagging: People-Powered Metadata for the Social Web,* New Riders Press, USA, 2008

Furthermore, in 2013, only 22 percent of the information in the digital universe would be a candidate for analysis, i.e., useful if it were tagged. (More often than not, we know little about the data, unless it is somehow characterized or tagged—a practice that results in metadata.) Less than five percent of that was actually analyzed. By 2020, the useful percentage could grow to more than 35 percent, mostly because of the growth of data from embedded systems.[15]

Given the statistics, this is a data gap that remains unexploited for enterprises to extract new insights or further refine existing conclusions. Tagging existing information will require a time investment to ensure proper categorization of the data, resulting in invaluable insights about consumers, etc., to optimize key business processes and uncover new monetization or learning opportunities.

1.2.11 Data Flows and Data Diagrams

A data flow diagram (DFD), also known as bubble chart or function model, is a representation of the flow of data through an information system in a graphic form. It is a model of the processes that transform information. DFD modeling is a primary tool during the SDLC design phase, given that it makes it easier to communicate technical information to a nontechnical audience to gather feedback. DFDs are also used for structured design modeling of data processing and indicate actors (i.e., users), expected input into and output from the system, data flow (i.e., data source and final resting place), and where the data will be stored.

DFDs usually do not include information about the timing of processes or information about whether processes will operate in sequence or in parallel.

DFD was conceptualized by two authors, Ed Yourdon and Larry Constantine, Massachusetts Institute of Technology (MIT) alumni and computer scientists with backgrounds in structured systems analysis and design. Their book, *Structured Design*, written in the 1970s, introduced DFDs and their uses. It built upon the data flow graph computation model work done by David Martin and Gerald Estrin. Other experts contributed to DFD notation: Trish Sarson, Tom DeMarco and Chris Gane.

DFD symbolism may present slightly different symbols depending on the authors followed, but the meanings are the same. Regardless of which is chosen, it is important to remain consistent.

[15] EMC Digital Universe, "Executive Summary: Data Growth, Business Opportunities, and the IT Imperatives," *The Digital Universe of Opportunities: Rich Data and the Increasing Value of the Internet of Things*, *www.emc.com/leadership/digital-universe/2014iview/executive-summary.htm*

Figure 1.6 depicts a standard DFD diagram.

Figure 1.6—Sample Standard Data Flow Diagram (DFD)

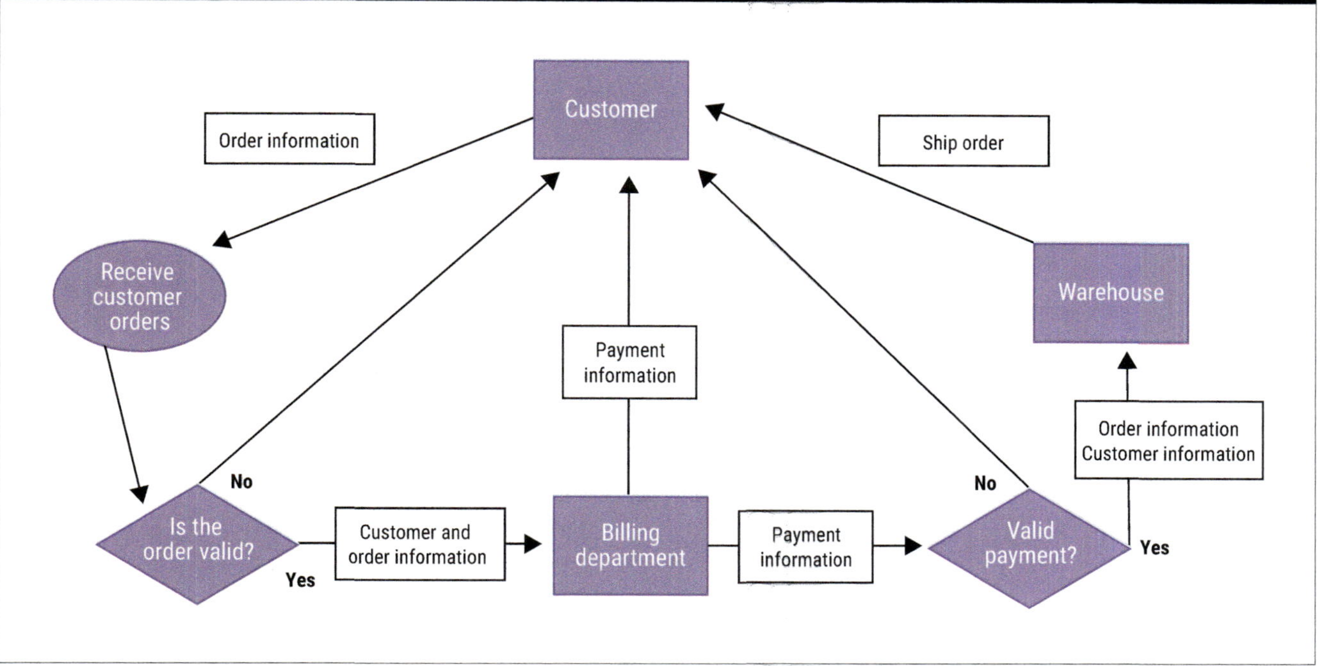

A DFD consists of four components: processes, flows, warehouses (i.e., data stores), and terminators that each have their own symbol.

- **Entities/sources**—Entities are end users or actors (i.e., individuals or a system) that provide input into the process but are outside the control of the DFD's boundaries. External entities provide data (i.e., input) that flows between the entity and the system being diagrammed. When the system described in the DFD provides these entities with data, they are called sinks, as the recipients of system output. An entity may either provide data (e.g., input from a scanner) or receive data (e.g., send to a printer). In this case the scanner and printer are not part of the system with which the DFD is concerned—they simply communicate with the system. The scanner is an entity and the printer is a sink.
- **Processes (what)**—Processes are activities that change or transform data flows. Given that processes transform incoming data into outgoing data, all must have inputs and outputs on a DFD. On a diagram this notation is given a name that correlates to its function, such as "Process order," rather than a complicated, technical name. Processes are typically shown from top to bottom and left to right on a DFD. Three sample process symbols are shown in **figure 1.7**, depicting DeMarco and Yourdon's notation. In Gane-Sarson notation, a box is used. It may be labeled with a requirement or function reference number, and a short title that describes its function.

Figure 1.7—Data Flow Diagram Using DeMarco and Yourdon's Notation

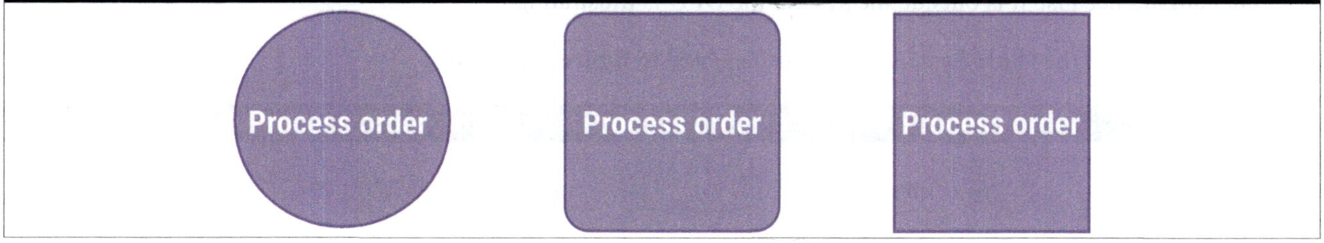

- **Flows (where)**—Movement of messages, impulses and signals between external entities, processes and data stores is represented with an arrow symbol, which indicates the direction of flow. This data could be electronic, written or verbal. Output and input and data flows are marked with the name written alongside the arrow.

  Figure 1.8 depicts the different combinations of flows.

Figure 1.8—Data Flow Diagram Samples

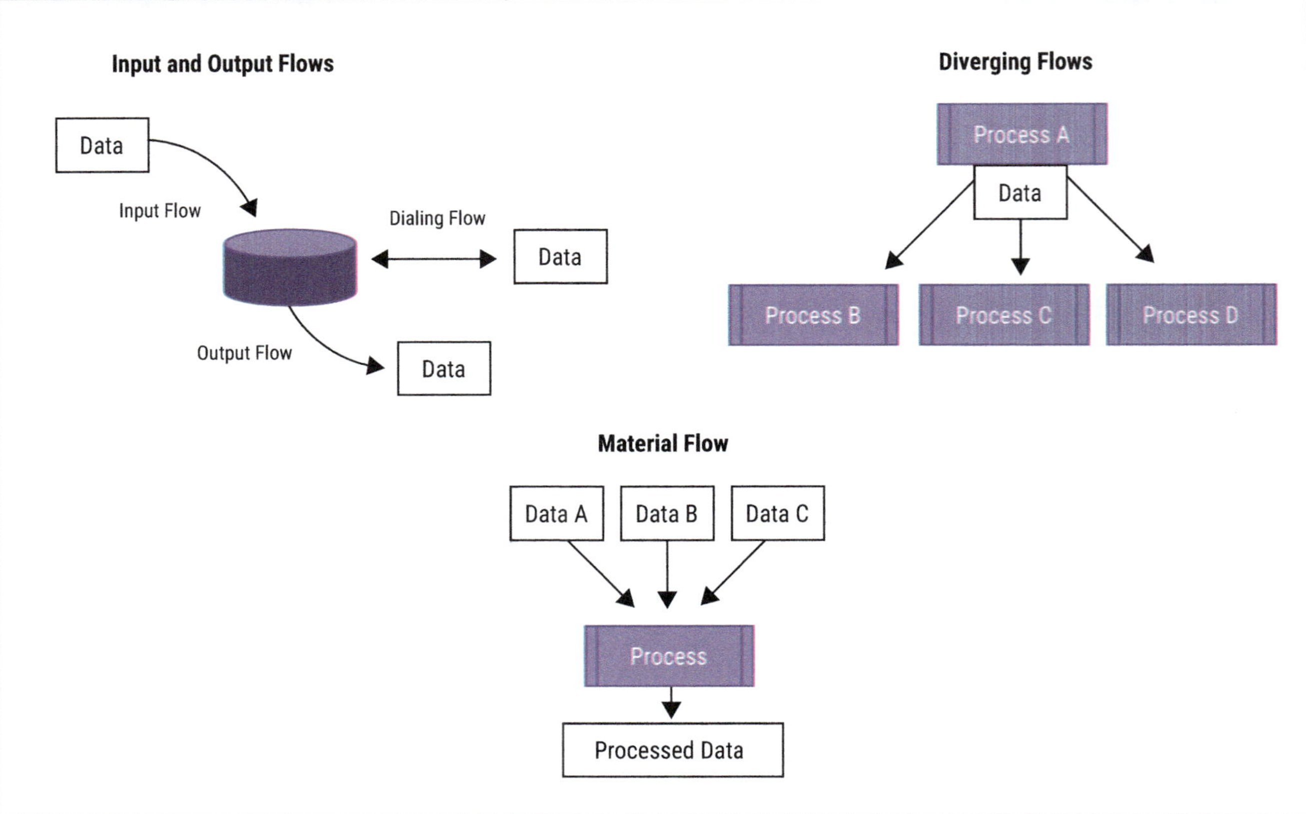

- **Warehouses**—Warehouses are also known as databases, storehouses or data resting places. A warehouse may take the form of electronic data storage, such as a flat file (.doc, .pptx, etc.) or database; or a hard copy, such as a printed PDF. A warehouse does not generate any operations but simply holds data for later access. Files can be stored long term or batch-stored briefly while waiting to be processed. Warehouses are used when data has to be retrieved sometime in the future.

  Figure 1.9 depicts the different combinations of stores.

Figure 1.9—Data Flow Diagram Data Store Notation

- **Terminators**—A terminator, which is notated using a rectangle, may be both a source of information and a receiver of information. It may refer to a department in an enterprise or a government entity. A terminator is unchanging because it is outside the system the DFD is illustrating.

  An overview of the Gane-Sarson notation is depicted in **figure 1.10** below.

Figure 1.10—Data Flow Diagram Using Gane-Sarson Notation

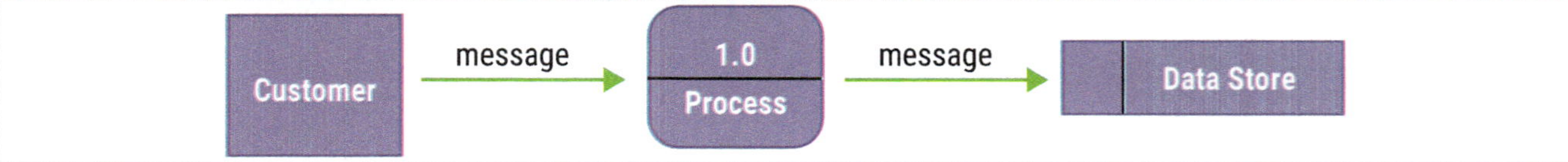

Although a DFD provides a well-though-out depiction of system processes and data transformations, it does not provide details about the components represented. To show details about *what* the information comprises and *how* the transformation occurs, a supporting document called a data dictionary usually accompanies a DFD.

A data dictionary stores all the details that correspond to the DFD stores, processes and flows. It may be a database that contains the name, type, range of values, source and authorization for access for each data element in a system. It indicates which application programs use those data, so that when a data structure is contemplated, a list of the affected programs can be generated. **Figure 1.11** shows a snippet of a sample data dictionary.

Figure 1.11—Data Dictionary Example
name = courtesy-title + first name + (middle name) + last name
courtesy-title = [Mr. I Miss I Mrs. I Ms. I Dr. I Prof.]
first-name = {legal character}
last-name = {legal character}
legal-character = [A-Z I a-z I ' I - I I]

A DFD has levels of complexity for the system (or requirements) illustrated. The levels typically are numbers from zero to two; in some cases a level three may be used, depending on the scope or depth illustrated. Level-0 is referred to as a context diagram and typically depicts a high-level overview of the system. Each subsequent level (e.g., level-1 and on) is more detailed, and is known as a decomposition level (i.e., level-1 DFD is more detailed than level-0, but less detailed than level-2).

In **figure 1.6**, a level-1 DFD may further elaborate on the details between the CUSTOMER AND ORDER INFORMATION and PAYMENT INFORMATION flow, for example, allowing a more focused examination.

Finally, a DFD must be balanced. The balancing concept dictates that in showing the next level of decomposition, every incoming and outgoing flow to a process in the preceding level must be preserved. This is to say that if RECEIVE CUSTOMER ORDERS has two inputs and three outputs in the level-0 DFD, it must retain them in level-1 diagramming. This is to ensure that as a subset of a system is observed, the higher contextual considerations are not lost.

Difference Between Data Flow Diagram and Flowchart

A flowchart presents a process in simple steps by displaying the steps in boxes connected by arrows. A flowchart details every step along the way, from the beginning of the process or flow to the end. Flowcharts can be used in any industry to chart business flows, network diagrams, or even an educational curriculum. DFDs are used mostly in the software development and computation industries.

1.2.12 Applicability of Data to Business

The data mining associated with the data science domain takes advantage of the computation power of computers to extract and analyze large data sets in search of patterns that assist enterprises in making business decisions.

Typically, data analysis in large enterprises entails extremely voluminous analysis of patterns and trends, which is extremely useful in ascertaining the efficiency and effectiveness of operations. Obtaining an understanding of data by considering samples, measurement and visualization helps enterprises deepen their understanding of consumer buying habits and demographics. Data analysis can result in laser-focused, effective and efficient product development and marketing campaigns. In the case of software development, it can identify features that will actually provide measurable value to the consumer and the enterprise.

In the context of competitiveness in the global enterprise market, business analytics (BA) and data science may appear to cross paths. They both involve data gathering, modeling, insight sharing and reporting. However, the difference between the two is that business analytics is specific to project-related concerns like cost, profit and market share, and it primarily uses past data with statistical tools and techniques.

Data science enhances BA tasks, tools and techniques by adding valuable dimensions:

- Asks and answers questions—e.g., on the influence of seasonal factors or geographical, political and socioeconomic trends—not only to analyze existing data, but also to predict behavior based on consumer preferences.
- Combines data gathering, analysis and mining with technology's best practices for algorithm building to answer a multitude of questions.

1.3 Data Structures

In software development, a data structure may be selected (built into programming language via keywords) or designed to store data for the purpose of manipulating it with various algorithms, and to organize code and information in a digital format. Each data structure contains data attributes (information), information on relationships between the data, and functions that can be applied to the data. Using the correct data structures assists with proper organization and retrieval of stored information.

1.3.1 Characteristics of Data Structures

Characteristics of data structures include:

- **Static or dynamic**—Static means fixed, unchanging. When data are processed and compiled, the fixed sizes, structures and memory locations of static data remain the same. Dynamic data structures have memory allocations that can shrink or expand depending on the input and processing at compile time.
- **Homogeneous or nonhomogeneous**—This trait indicates whether all data items are of the same type or of various types (e.g., integers or floats).
- **Linear or nonlinear**—Data can be ordered (i.e., chronological order, array) or non-ordered (e.g., a graph).

Figure 1.12 shows an overview of data structures.

Figure 1.12—Data Structure Types

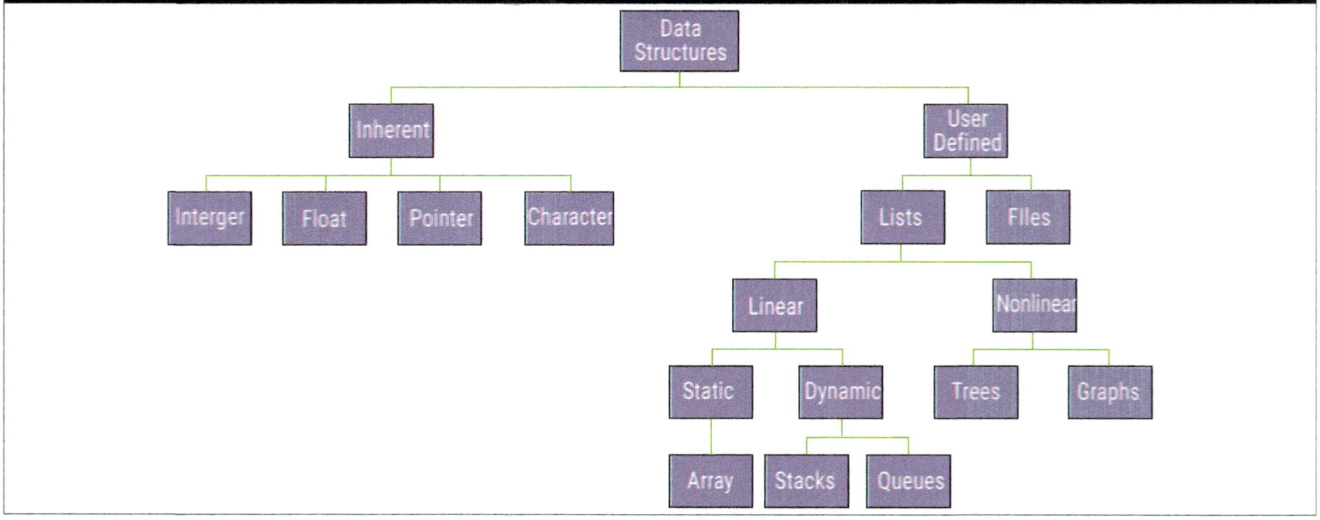

Primitive (i.e., inherent, built-in) and advanced (i.e., user-defined) data types:

- **Integer**—Integers are mathematical whole numbers, commonly represented in a computer as a group of binary digits (bits). For example, 21, 4, 0 and -2048 are integers, while 9.75, 5+1/2 and √2 are not.
- **String**—A string is a series or a set of characters (i.e., char) represented as a single piece of data, most often notated with double quotes. The string may be constant (fixed, unchanging) or variable (changing).
- **Character (char)**—The character data type is a single digit string, unlike the string data type, which can allot for many more characters.

"Hello World" String	"A" Character string	"Hello John" Variable string <name>

- **Boolean**—The Boolean data type allows one of two possible values: 1 or 0, which correlate to TRUE/FALSE or ON/OFF, respectively.
- **Floating-point number (float is 7 and double is 15)**—The floating-point data type, which is used to represent fractional values, applies to decimal numbers, such as 19.95 or 3.141578907654321 (decimal digits of precision). Decimal point precision requires use of the right float, especially in the financial sector where every bit adds to a sum balance. The wrong float can result in a significant error.

To understand advanced data structures, it is important to note that *how* and *when* data are updated can make a difference, because the data structure sometimes determines the ability to update data via two concurrent processes (i.e., multiple input streams, multi-threaded). Each time a program executes a statement of code it starts a process (or multiple processes) called a thread.

Thread-safe code used for reading or writing manipulates shared data structures in a manner that safeguards all other threads without unintended interaction. There are several strategies for making thread-safe data structures. A developer may opt to execute code in several threads simultaneously in a shared memory space where each of those threads has access to all of the addresses of every other thread.

Thread safety is a feature that allows code to run in multi-threaded environments by means of inherent or created synchronization to reestablish some of the correspondences between the actual flow of control and the text of the program. Synchronous and asynchronous processing are terms used to describe *when* a program executes the input instruction set.

Synchronous processing executes at specific time intervals or when another process the program is dependent on has completed. It can block other processing and create bottlenecks.

Asynchronous processing is independent of the timing requirement and can execute multiple instruction sets concurrently. Different actions are executed at the same time without the need to wait for another to complete.

1.3.2 Linear Structures

Arrays can be either static or dynamic and are used to store multiple data items as a grouping or a collection using neighboring memory locations. Items stored this way are usually the same; hence the position of each element can be calculated or retrieved easily.

Figure 1.13 shows a graphical representation of an array structure.

Figure 1.13—One, Two and Multi-Array Structures

One-dimensional array

Index	1	2	3	4	5	6	7
Value	A	B	F	G	T	Z	Y

Two-dimensional array

Index	1	2	3	4
1	A	B	F	G
2	X	Y	L	T
3	C	E	H	D

Multidimensional array

Index	1	2	3	4
1	A	B	F	G
2	X	Y	L	T
3	C	E	H	D
3	J	I	K	M
3	O	Q	U	V

A stack stores a collection of items in linear order (stacked one on top of another in order of receipt). Order can be first in first out (FIFO) or last in first out (LIFO) to describe the way the data are replaced or overwritten.

Figure 1.14 shows a graphical representation of a stack structure.

Figure 1.14—Stack Data Structure

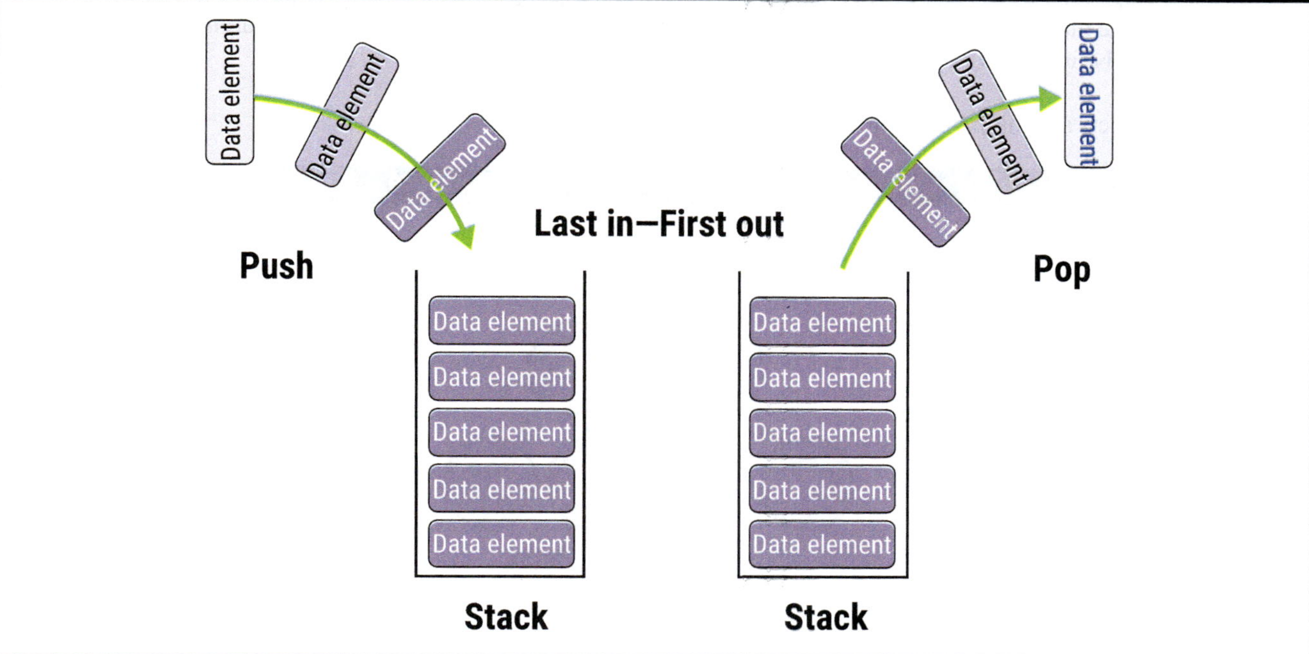

Source: Voronyansky, Y.; "Java Data Structures," DEV, 13 January 2019, *https://dev.to/vrnsky/java-data-structures-417e*

A queue stores a collection of items in a manner similar to a stack. However, the operation order can only be first in first out (FIFO) and is typically used when priority is important. A stack has a top concept, while a queue considers front and rear when making decisions. For example, a call center takes the calls in the order they come in—first to call in is first to be helped.

Figure 1.15 shows a graphical representation of a queue structure.

Figure 1.15—Queue Data Structure

A linked list stores a grouping of items in linear order. Each element, or node, in a linked list contains a data item as well as a reference, or link, to the next item in the list.

Figure 1.16 depicts a sample linked list data structure.

Figure 1.16—Linked Data Structure

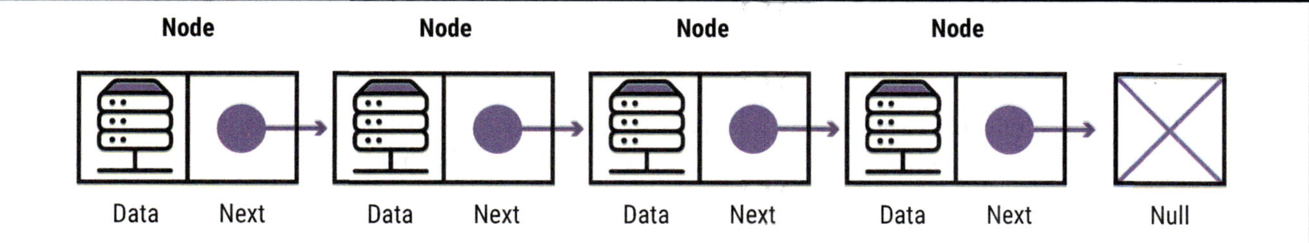

1.3.3 Tree Structures

As the name implies, a tree is an abstract hierarchical view that highlights data relationships via nodes, also known as children. A tree may be binary (i.e., 1, 0) or heap. A binary tree has a maximum of two children connected by edges (i.e., lines between nodes).

Figure 1.17 depicts a sample binary tree with its components.

Figure 1.17—Binary Tree Data Structure

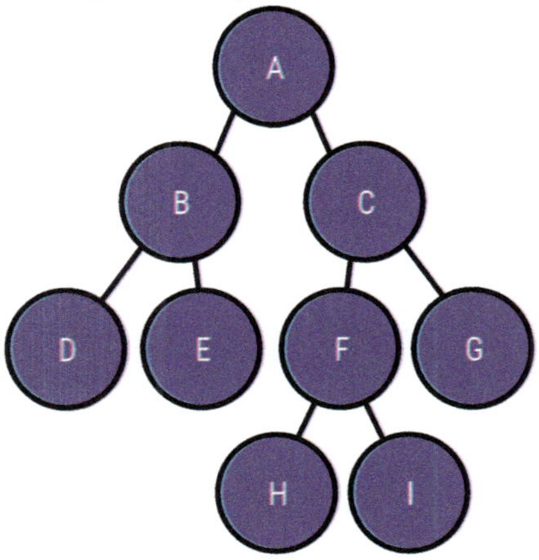

A heap tree is a composed binary tree data structure in which the root-node key is compared with its children and arranged according to either min-heap or max-heap. A heap tree described mathematically:

If α has child node b then $-$ key(α) $\geq$ key(b)

Figure 1.18 depicts binary min-heap and max-heap trees.

Figure 1.18—Min-Heap and Max-Heap Binary Trees

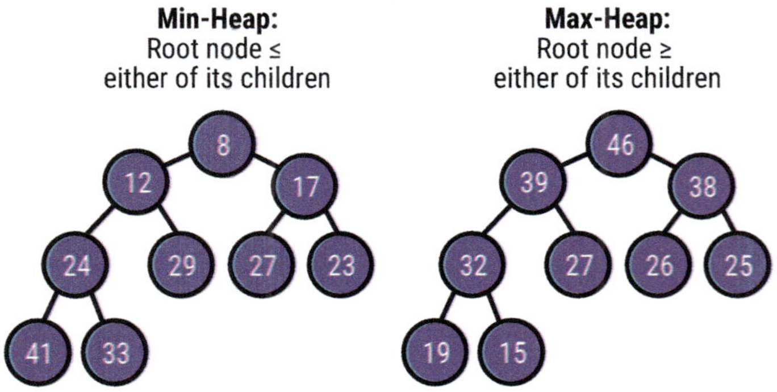

In contrast with tree, which is binary, the trie structure is used for strings. "Trie" is part of the word retrieval because this data structure excels at fast pattern matching, enabling quicker retrieval.

1.3.4 Index and Pointer Structures

A hash table is a map, a schematic showing where data is stored. It uses an associative array that charts keys to values, and a hash function to convert an index into an array of containers that hold the desired data. Although the hash table/hash map structures are the same (they both point to an interface that maps keys to values), each has slightly different functionality.

A hash table is slower, but it is synchronized and thread safe and can be shared with other threads, whereas a hash map is not thread safe and will require additional synchronization code.

Figure 1.19 shows a sample hash table/map.

Figure 1.19—Hash Table / Map Data Structures

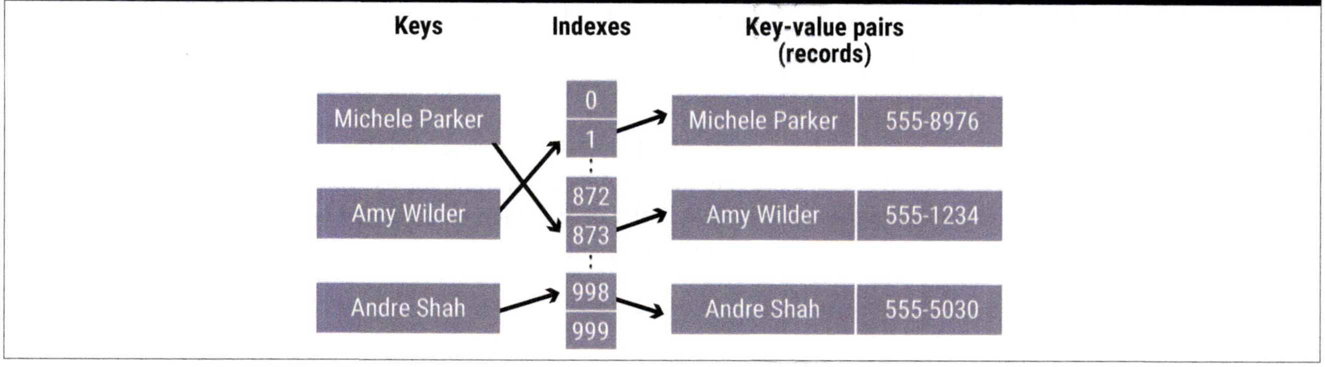

Graphs are useful for representing any network (i.e., web) type of data, such as devices on a network. The data stored in graphs is not ordered, in contrast with stacks. Graphs are made up of a predetermined set of nodes (i.e., vertices) and lines that connect them (i.e., edges) and are considered nonlinear and nonhierarchical with no root node.

Figure 1.20 shows a sample graph structure.

Figure 1.20—Graph Data Structure

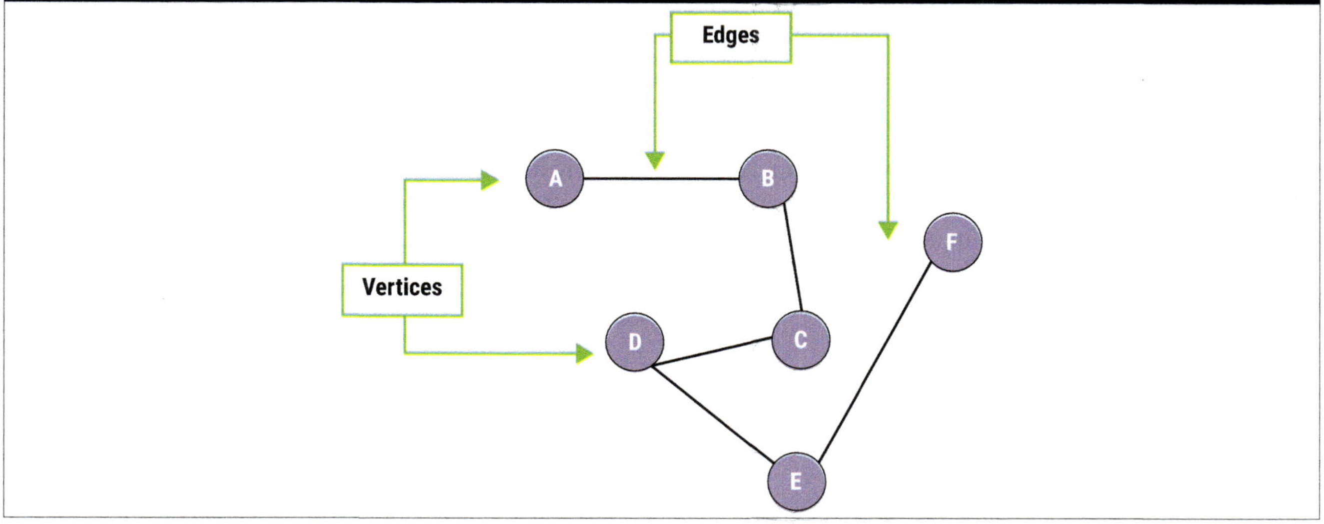

Indexes

An index can be compared to a book index (usually found in the back of the book) that includes keywords and page numbers to provide quick access to pages that discuss a particular term or subject, instead of requiring them to look through hundreds of pages to find all the references to that term. Another example may be a list of home addresses associated with individuals and organized and ordered in a phone book by their last names. This idea is also applied programmatically in computer science for data stores, and specifically in big data to improve the speed of information retrieval (also known as LOOKUP in database lingo) from the memory locations where the information resides.

There are several types of index implementations, and they vary depending on constraints such as data models, data types, content, and relationships between the tables. Indexes can be created by copying specific columns of tables, e.g., phone number area codes or individuals' last names. (Which would more likely be used as search criteria?) They can be implemented based on functions or as partial indices based only on specific, unique records of data.

Finding information quickly is very important—hence the different types of indexes one can implement to meet speed benchmarks while working with existing constraints and reducing the processing overhead required. Sequential search, also known as linear search, is the slowest and the most accurate for voluminous datasets, as it reads and iterates through every record in sequence until the record is found or the end of the data set is reached.

An indexing application must consider the data model. Further, the index type coupled with the data model dictates the query that will be the most effective. For example, the candidate columns for use in a non-clustered index key are the ones that are frequently involved in the GROUP BY, JOIN or WHERE clauses that will cover submitted queries and return exact match values rather than larger, less precise results.

Pointers

A pointer references a location in memory and obtains the value at that location. The process of obtaining the value stored at a location referenced by a pointer is known as dereferencing. A reference pointer can be implemented to point to a specific memory location. It removes redundancy of duplicated data in multiple memory locations and expedites search. See **figure 1.19**.

Pointers in data structures can be arithmetically manipulated via code. They can be incremented or decremented by adding or subtracting the values found in the prior or latter point, respectively. Some cautions:

- Dereferencing a pointer with an invalid memory location can result in a complete application crash.
- Because pointers allow protected and unprotected access to memory locations, ensuring they are valid can guard against unsafe code.

In languages such as Python, PHP, JavaScript or Java, pointers play an important role behind the scenes in the form of references to objects. When an object variable is used in a Java program, the computer automatically follows the pointer stored in that variable in order to find the actual object in memory. In Python, however, pointers are replaced by mutable objects—those whose value can change after they are created.

1.4 Statistical Analysis

1.4.1 Populations and Samples

Statistical sampling is a method of selecting a portion of a population, by means of mathematical calculations and probabilities, for the purpose of drawing scientifically and mathematically sound inferences regarding the characteristics of the entire population.

A statistical population is a set of concepts, people, animals, species or any components that share defined properties from which a sample is drawn. Population sampling is one of several statistical techniques used for analysis. Statisticians identify a certain population (i.e., target population) representative of the properties they'd like to observe and then seek answers to some questions. For example, they may identify a set of 300 individuals over the age of 40 with income ranges of $40,000-$65,000 (elements or parameters to be studied) and attempt to gain insights into their buying habits, or discover what percentage of income they spend on a dwelling. Each individual in this case is known to be part of a statistical frame.

Those 300 individuals, based on their respective attributes, were chosen to represent the population for the purpose of conducting the analysis. Comparing the size of the statistical sample to the size of the population produces a sampling fraction. In sampling theory, the sampling fraction is the ratio of sample size to population size. Having very large numbers of sampling population can increase research and analysis time and costs. Considering the subject matter of data science, the size of the population sample data is critical, because the computation power and resources required for very large samples may introduce technical complexities not seen in smaller sample sizes.

Figure 1.21 shows a visualization of the population sampling technique.

Figure 1.21—Population Sampling Technique

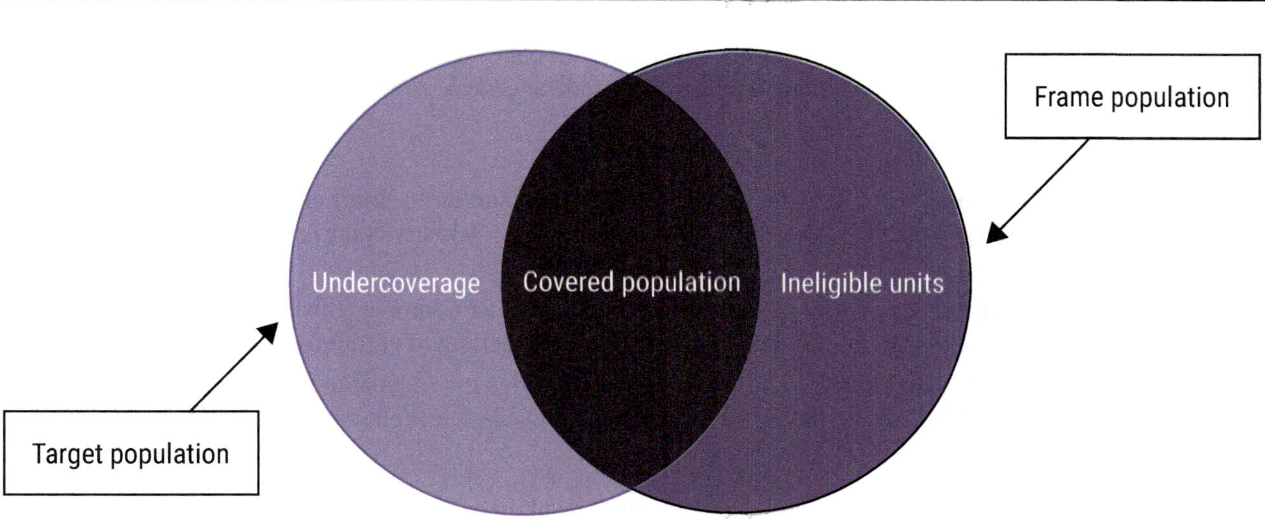

A subpopulation is a refinement (i.e., reduction) of an existing population sample. In the example above, the analyst could reduce the population sample to identify all subjects with the designated properties (over 40 years old, income range $40,000-$65,000) who were associated with the dwelling type of townhomes.

Probability Sampling

Probability sampling involves taking a random selection of samples from the greater population. It ensures that each entity of the population has an equal chance of being selected. It is the least costly, and in addition to saving time, it offers the best chance to create a sample that is truly representative of the population.

Types of probability sampling:

- **Simple**—Each entity of the population has an equal, or non-zero, chance of being selected.
- **Systematic (interval)**—Entities are selected according to a defined orderly approach (e.g., every third car is selected).
- **Stratified**—Entitles are classified and organized into subgroups (i.e., **strata**) based on defined parameters to make the process of selecting a representative sample with those parameters more efficient and accurate, and to account for all subgroupings. However, the stratified sampling approach is not the best option when there are no homogeneous subgroups (e.g., there are many colors, but not enough of each). Also, it is a more expensive approach compared to other methods.

- **Cluster**—The population is divided into clusters, often with wide-scale parameters, such as geography or time (e.g., all individuals born in the 18th century). Within each cluster there likely are randomized entities that would be selected if that cluster were selected and then further refined based on other variables.
  - **K-means clustering**—A data mining algorithm to cluster, classify or group N objects based on their attributes or features into K number of groups (so-called clusters). K-means is discussed further in section 3.4.2.
  - **K-nearest neighbors (K-NN)**—A machine learning algorithm that classifies things based on their similarity to nearby neighbors. The algorithm's execution is tuned by picking the number of neighbors to examine (K) as well as some notion of "distance" to indicate how near the neighbors are. K-NN is discussed further in section 3.5.6.

Non-Probability Sampling

Non-probability sampling entities do not have an equal chance of being selected. Therefore, there is an increased risk of ending up with a nonrepresentative sample that may not produce generalizable results.

Types of non-probability sampling:

- **Quota**—Divides the sample into mutually exclusive subgroups, similar to stratified sampling. Then selection is made from each of the groups. (e.g., all colors are subgrouped into primary color families and then a count from each is selected.)
- **Convenience (accidental, opportunity)**—The fastest way to gather a sample, either because the entities are easy to reach or they are readily available in the population and willing to participate.
- **Judgment (purposive)**—The sample is chosen based on the judgment of an authority figure or regulating body. This method is fallible, because it relies on personal judgment. The selection may be tainted by a myriad of biases.

In computer programming and statistics alike, garbage in will result in garbage out. Taking the time to choose the correct statistical technique and implementation will lead to a more accurate analysis and higher-quality results.

1.4.2 Statistical Modeling

Statistical modeling is a combination of interpretations, based on collected data and population understanding, used to predict information (see **figure 1.22**). Visual and graphical representations are used to convey analysis results and conclusions in an easily understood way. Visual models can be simple flowcharts, pictures or infographs, or they can take the form of mathematical equations or computer models.

Several types of models are used as tests to find the relationships between data variables. For example, if sample data reveals that most individuals in the $65,000 salary range are homeowners and those at the $40,000 range are renters, a correlation could be inferred.

A model is useful if it:

1. Is not overly complex and hard to understand, thus negating its purpose
2. Describes the research and sampling techniques used (i.e., how the data was collected)
3. Highlights the questions it aims to answer
4. Describes the inferences and relationships uncovered (i.e., conclusions, key takeaways)

Figure 1.22—Statistical Modeling Overview

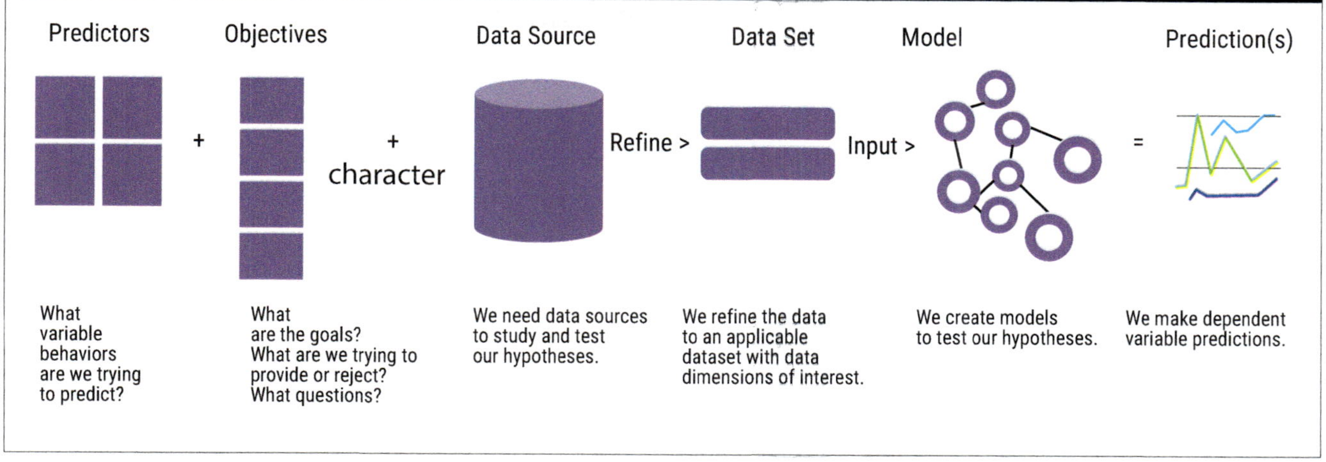

There are two main categories: regression models and classification models.

Regression Models

Data analysts use regression models to deduce relationships between variables. Regression models are often used to uncover which independent variables (e.g., salary range) are the most influential over the dependent variables (e.g., spending money)—valuable information that enterprises can leverage to make essential business decisions.

Numerous types of regressions can be performed. The traditional regression models are logistic regression, linear regression and polynomial regression. Examples of other regression models include stepwise regression, ridge regression, lasso regression and elastic net regression. Each method has value and is the best choice to apply to specific conditions in the data science field. In the most common, linear regression, the analyst finds the line that most closely fits the data, according to a specific mathematical condition.

- **Linear regression model**—Used to predict a target variable by fitting the best linear relationship between the dependent and independent variable (see **figure 1.23**). Simple linear regression predicts a dependent variable by using a single independent variable and assumes there is a relationship between the variables.
  - **Purpose**—It is utilized for identifying the relationship (or predicting it) between a dependent and independent variable.
  - **Formula**—$Y= b_0 + b_1 * x_1$

    Y denotes the dependent variable (DV) or variable to be predicted, b_0 is the change in Y relative to a one unit change in x_1 holding all other independent variables constant (i.e., when the independent variables are held at the same value or are fixed). B_1 is the slope, and x denotes independent variable (IV).

Example: Several data points of homes' prices and their square footage are plotted in **figure 1.23**. The purpose is to see if the home's prices (i.e., Y dependent variable) are dependent upon the square footage of the home (i.e., X independent variable).

Figure 1.23—Linear Regression Using Scatter Chart

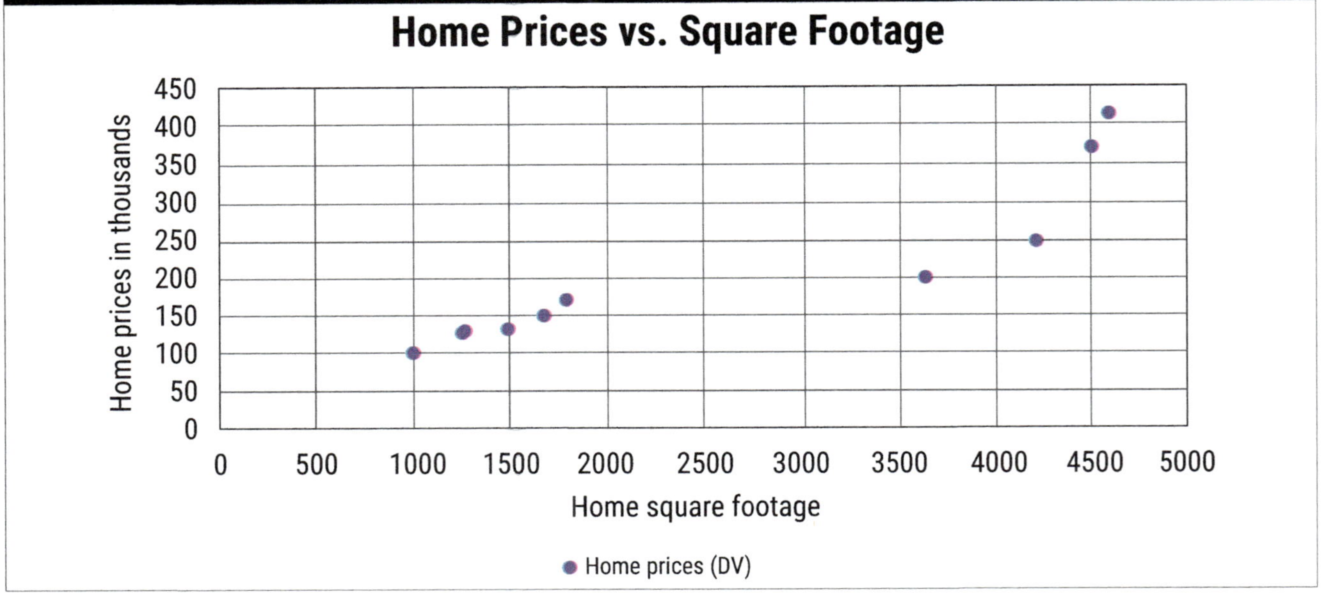

Figure 1.24 shows the commonly used simple linear regression model using a scatter chart to show a correlation between homes square footage and their sales prices. The lower right of the image shows the formula: Y = Home Prices, x = the square footage slope (0.066), the intercept (34.458), and R^2 denotes regression squared. R^2, the sum of the squares of the distance of points from the line, can range from 0 percent-100 percent. In this case it is 0.8339, signifying a positive linear association. The linear regression model's score (R^2) shows how well the model fits the data, i.e., does the model explain the changes in home prices well? Are home prices affected by square footage of the home?

Figure 1.24—Simple Linear Regression

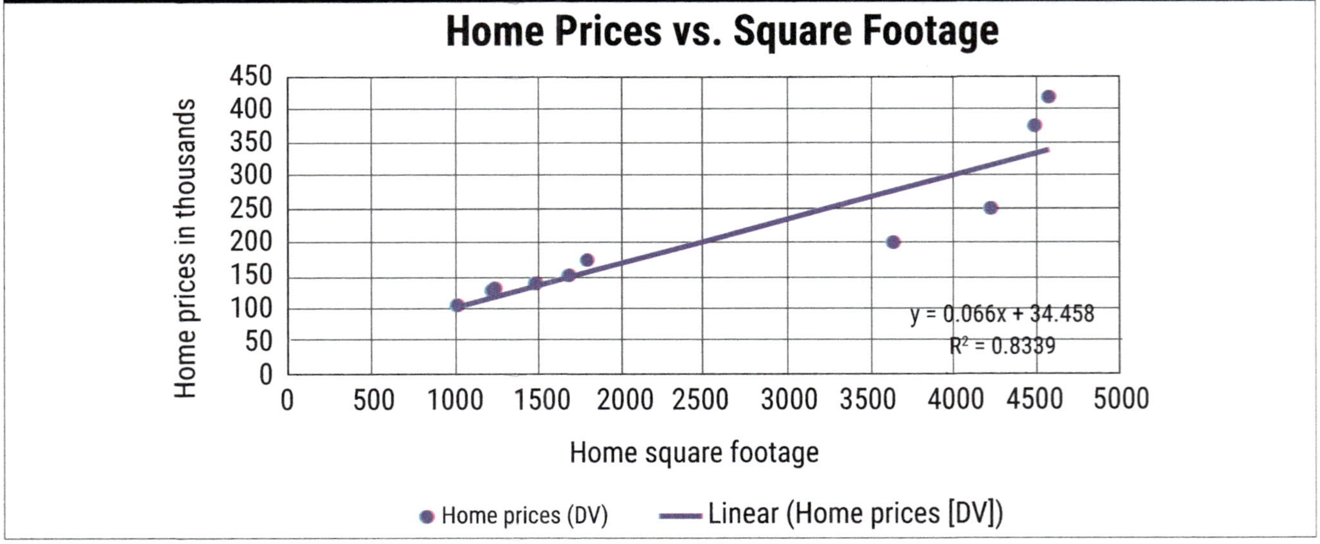

- **Multiple Linear Regression**—Reveals relationships between a dependent variable and two or more independent variables, i.e., it allows the analyst to estimate the association between the independent variable and all other potential confounding variables.
  - **Purpose**—Utilized for prediction, theory building or explanation of a hypothesis.
  - **Formula**—$Y= b_0 + b_1 * x_1 + b_2 * x_2$

Y denotes the dependent variable (DV), b0 is the change in Y relative to a one unit change in x1 holding all other independent variables constant (i.e., when the remaining independent variables are held at the same value or are fixed). There should be one regression coefficient (e.g., b1) for each independent variable (i.e., x1).

Example: Using the dependent variable of home prices and two independent variables of age of home, and square footage, the resulting regression value of 0.23779 signifies there is a weaker relationship (i.e., based on this data, the age of the home does not affect its pricing substantially).

Figure 1.25 shows an example of the multiple linear regression model comparing the dependent variable of home prices to the two independent variables of square footage and age of home. **Figure 1.24** identifies a positive correlation between square footage and home prices. How about age of home and home prices?

Figure 1.25—Multiple Linear Regression

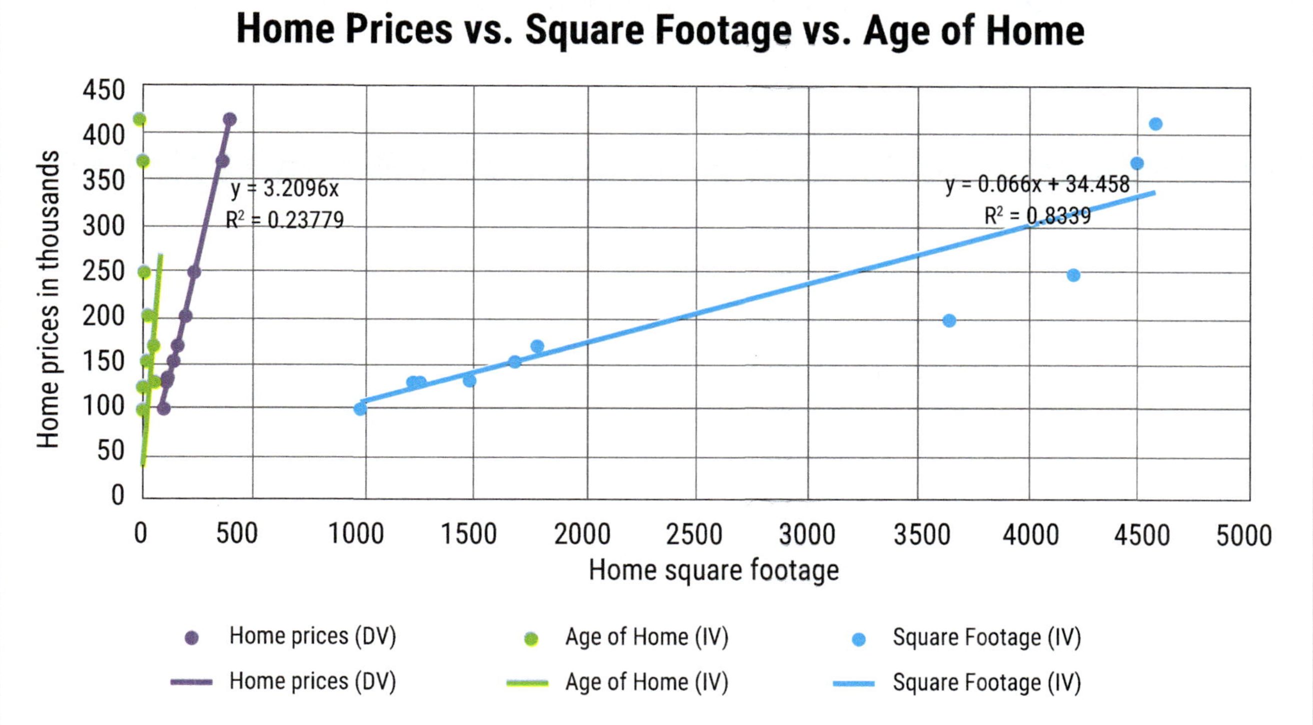

There are three different ways of building regression models:

- **Simultaneous**—All independent variables are entered together (i.e., tested together). This adds complexity and could create a GIGO scenario.
- **Stepwise**—Independent variables are added (forward stepwise) or removed (backward stepwise) one at a time by size of data, correlation or a predefined order.
- **Hierarchical**—Independent variables are added in stages.

Classification Models

Data mining is used by data scientists to find usable models and insights in data sets. Numerous techniques are employed to accomplish this task, such as outlier analysis, regression, cluster analysis and classification.

Classification is a data mining technique that uses a programmatic algorithm to analyze an existing data set of known points. It assigns categories to data collections to improve the accuracy of predictions and analysis.

Classification is a form of machine learning that can be particularly helpful in analyzing very large, complex sets of data to lead to more accurate predictions. Classification is applied to both structured and unstructured data, although the latter is more difficult to classify and the process is error prone. Unstructured data is any data that does not fit a predefined data model. Often this data does not fit into the typical row-column structure of a database. Images, emails, videos and audio are examples of unstructured data.

Classification models are a form of supervised machine learning that is often used when analysts need to understand how they arrived at a certain point. They provide more than an output. They provide additional information that can be used to explain the results of the prediction to project stakeholders.

Some of the most common classification algorithm models include decision trees, K-nearest neighbor, logistic regression, random forests, Stochastic Gradient Descent, Support Vector Machine (SVM) and Naïve Bayes. A decision tree (**figure 1.26**) is used to visually describe the decisions and decision-making route of the analysis, and each of the nodes (i.e., branches) has its own probability. The decision trees can be used to drive informal discussion of possibilities further refining the questions that need to be asked, or used to map out an algorithm that predicts the best choice mathematically.

Figure 1.26—Decision Tree

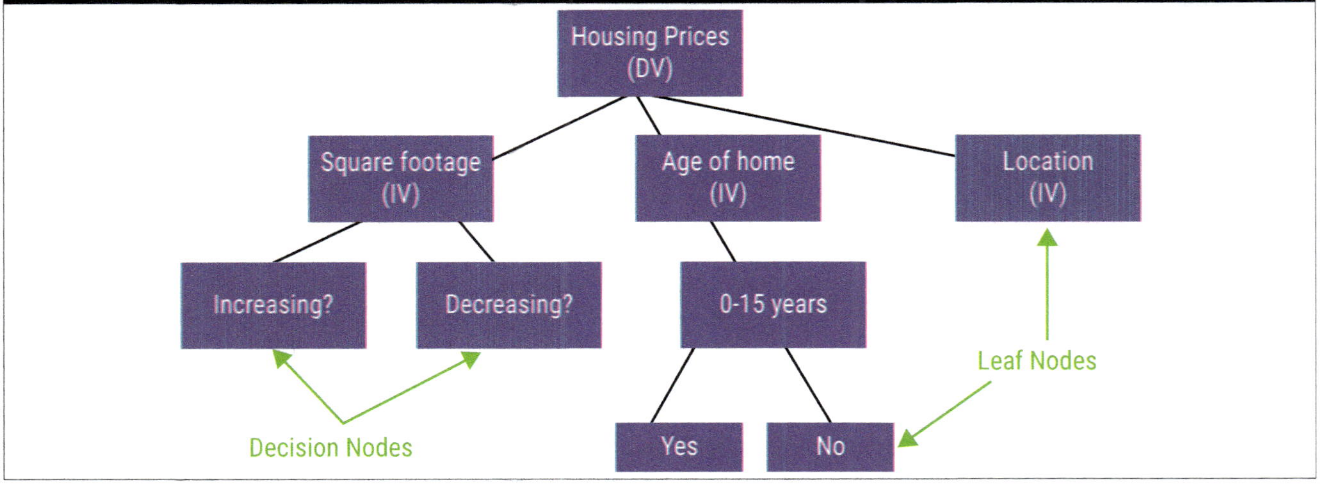

A decision tree is incomplete if the decision nodes have not been exhausted into leaf nodes (as is the case in **figure 1.26**). Classification rules require the inclusion of all test scenarios (decision nodes) and all class variables in the final output. For example:

- If the square footage is increasing, does it affect housing prices?
- If the square footage is decreasing, does it affect housing prices?
- If the age of the home is between 0-15 years does it affect pricing? What about older homes?

Classification is also an example of pattern recognition. Using machine learning terminology, classification is an example of supervised learning. Supervised machine learning occurs when the input data has been labeled and tagged clearly to "assist" the learning algorithm in inferring the result. The supervised algorithm is taught to classify input into specific known classes based on prior, already classified, examples.

Conversely, unsupervised machine learning uses unlabeled data. The algorithm groups data (i.e., clusters data) into categories based on some predefined measure of inherent similarity or distance proximity (i.e., neighborly, e.g., pink is a neighbor of red). The algorithm makes an educated guess based on already-labeled data. Typically, independent observations are analyzed to determine a set of quantifiable properties, known variously as explanatory variables or features.

Neural networking models are also used, but they are reserved mostly for artificial intelligence modeling. They offer accurate predictions and can be very powerful and complex, but they typically are not helpful when explaining what is happening behind the scenes.

More regarding this unknown (i.e., black box) processing: This process can be likened to making cookies. The inputs may be the ingredients, which are known (e.g., chocolate chips, dough), and the output is freshly baked chocolate chip cookies. The process of how they were created is unknown. In some cases, an analyst who is familiar with the inputs and the outputs can make educated assumptions about the process of data transformation that occurred (i.e., how chocolate chips and dough transformed into cookies). In this example, mix all ingredients, set oven to 375°, etc., would be a fair assumption, but the exact mechanism of how it happened was hidden.

1.4.3 Key Performance Indicators (KPIs)

A key performance indicator (KPI) is a measure that determines how well a process is performing toward a goal to be achieved. A key goal indicator is defined as what has to be accomplished (i.e., the final outcome). In comparison, a key performance indicator is a measure of how well the process is performing. In other words, KPIs provide insight into business conditions like product quality, predictability, return on investment (ROI), and much more.

Goals that are agreed upon by all relevant stakeholders ensure a consistent progression toward improvement, and KPIs allow analysts to measure progress and optimize toward a single end result. To create KPIs, it is important for the development and analysis groups to work closely with the business to identify goals and determine how they will be measured and managed. A very rudimentary example is the goal to lose weight (i.e., end goal). To measure progression toward that goal, a person might include key performance metrics such as weight, fat percentage, etc., and set a target weight and fat percentage.

To be true indicators of progress toward a desired goal, KPIs must have the following characteristics:

1. Tied to a worthy goal worth achieving (i.e., not arbitrary)
2. Show a direct relationship to a business, personal, process, department, etc., goal
3. Be capable of measurement (i.e., quantifiable) via a set target (e.g., max one percent error margin)
4. Be actionable, capable of being affected (i.e., controllable)

Setting KPIs at an enterprise or departmental level requires the involvement of more than one individual. Transparent, constant communications are recommended, as KPIs may need to be changed, updated or completely revamped as more information becomes available and priorities change.

Lag indicators are key measurements that articulate the current state of the item to be measured. These are also the benchmarks—the starting point before any effective action is taken (e.g., current customer service call volume of 3000 calls per day).

Lead indicators are actions that can be taken to predict a more desirable future condition so enterprises can achieve goal forecasts (e.g., add Frequently Asked Questions (FAQs) to website and incorporate product testing to reduce daily call volume by 45 percent).

Balance score card (BSC) is concerned with four distinct areas—customers, finances, internal processes, and learning and growth perspectives. KPIs that are tied to any of these four pillars are much more welcome and better supported in an enterprise (e.g., heavy call volumes resulting in frustrated customers and higher employee costs for additional call representatives).

BSC is common across most mature enterprises with shareholders. It balances the needs of each of the four pillars in a factual, measurable performance approach. The four dimensions can be summarized as follows:

- **Customer**—The organization defines the business units' segments and areas of market competition. Performance metrics may include customer service, retention, satisfaction and new customer acquisition.
- **Finance**—All spending, cash flow and profits of the prior year are reviewed, and new goals are set using several calculations, including return on investment (ROI), before additional funds are allocated or business action is taken. Assets and liabilities are captured and assessed to determine if changes need to be applied. Sales forecasts and new product launches are subjected to in-depth analyses.

Figure 1.27 shows the ROI formula.

Figure 1.27—Return on Investment (ROI) Formula

$$\text{ROI} = \frac{\text{Profit (revenue from investment - cost of investment)}}{\text{Investment}} * 100$$

An example may be a business project tasked with implementing a new application subscription and corresponding services to improve a customer data platform's targeting. The project's total initial cost is $100,000 for the resources and the services needed. The added profit of the new customer acquisitions gained due to the project is $300,000. As **figure 1.28** shows, this business decision resulted in a 200 percent ROI; for every $1 invested, the organization gained $2.

Figure 1.28—Return on Investment (ROI) Example

$$200\% = \frac{200{,}000\ (300{,}000 - 100{,}000)}{100{,}000} * 100$$

- **Internal processes**—Enterprises are always interested in reducing costs, improving processes, and optimizing the way they do business. This perspective's outcome affects every other dimension in the BSC, because each internal process is reviewed for quality, efficiency and ROI.
- **Learning and growth**—The enterprise provides its biggest resource—people—with opportunities for continuous education and learning to build competencies, which in turn benefits the enterprise.

Implementing KPIs without a thorough analysis of lag indicators can skew the end goal and lead indicators. More importantly, not taking the proposed actions indicated by lead indicators may result in ineffective KPIs that reinforce the status quo.

A well-known KPI is the Net Promoter Score (NPS), also known as Net Performance Score, which measures customer loyalty to the brand in question. Customers are asked how likely they are to recommend a product or company to their friends and family, and their responses are grouped into promoter (9-10), passive (7-8), or detractor (0-6) categories. The score is then calculated by percent of promoters minus percent of detractors.

1.5 Chapter 1 Knowledge Check

REVIEW QUESTIONS

1. Choose the **BEST** answer. Which of the following is needed for data to have meaning?

   A. Data integrity
   B. Quantifiability
   C. Connectivity

2. The purpose of *wisdom* in the DIKW pyramid is to:

   A. Receive and disseminate information
   B. Transform data into meaningful, actionable results
   C. Add context to raw data

3. Which of the following is the **BEST** use of staged observation?

   A. To test hypotheses in instances where organic data are not readily available
   B. To amass large data sets for processing
   C. To expedite statistical analysis

4. Flows symbolism in a DFD depicts which of the following?

   A. Entities external to the system
   B. Movement of messages
   C. Both a source and a receiver of information

5. Static data structures characteristic means:

   A. Data are changeable and memory can either shrink or expand.
   B. Data structure is unchanging.
   C. Data structure is homogeneous.

Answers on page 46

Chapter 1 ANSWER KEY

Review Questions

1. A. Data integrity
   B. Quantifiability
   C. Connectivity. Refer to page 18.

2. A. Receive and disseminate information
   B. Transform data into meaningful, actionable results. Refer to pages 22-24.
   C. Add context to raw data

3. **A. To test hypotheses in instances where organic data are not readily available. Refer to page 24.**
   B. To amass large data sets for processing
   C. To expedite statistical analysis

4. A. Entities external to the system
   B. Movement of messages. Refer to page 27.
   C. Both a source and a receiver of information

5. A. Data are changeable and memory can either shrink or expand.
   B. Data structure is unchanging. Refer to page 30.
   C. Data structure is homogeneous.

Chapter 2:

Data Management

Data Management

Data Management

2.1 Learning Objectives

After completing this chapter, learners will be able to:

1. Distinguish among different data storage and management systems.
2. Describe the benefits of using automated processes to manage data.
3. Identify elements within a database management system.
4. Explain the use of data in online and cloud-based applications.
5. Detail data governance roles and responsibilities.
6. Explain legal, regulatory and ethical considerations regarding use of data.
7. Distinguish among data obfuscation, tokenization and encryption.

2.2 Data Management Systems

2.2.1 Introduction

Any large collection of information requires storage of some kind for safekeeping and later retrieval. An enterprise that uses numerous paper documents most likely will store them in large file cabinets. For an enterprise that primarily uses electronic data, the best storage options are likely large hard drives and almost unlimited cloud storage. For the voluminous data and associated metadata used in data science, consideration of the data management system to use is very important.

Big data, a term often referenced in data science, originated from the Moore's law concept that computing power (e.g., speed and memory) increases by a factor of two every two years. If it is linear, it means that information also doubles exponentially every two years. This notion has supported the need for careful consideration of storage, because massive data sets are created every day by millions of computational devices. Big data will not fit in a single computer or server. Conventional tools like Excel and SQL are typically unable to handle big data structures, so new ones have been developed to take their place. Moreover, a simple computer file structure does not provide the accuracy, availability, retrievability, optimal organization or transformational power required in the data science field.

2.2.2 Operational Databases

Operational databases provide data updates in real time. Real-time processing of data either takes place immediately or appears immediate to the user. The handling is done as the data is entered, so a continuous stream of input data is required to provide a continuous output. A good example is a well-known search engine named Google: As a user enters a search term, results appear to finish the user's question almost instantaneously.

An operational database (also known as online transaction processing, or OLTP) can store data used for day-to-day enterprise operations. OLTP systems are built for quicker access and ability to update customer information, purchase data and other transactional data. Another example is inventory systems that are tied to the registers. As soon as an item is purchased at the register, the item is removed from available stock.

A key benefit of an operational database is that it supports high-volume low-latency access, which enables it to perform the following functions in real-time: create, read, update and delete (CRUD, a computer acronym) and differs from conventional databases that perform updates in batches.

2.2.3 Relational vs. Non-Relational Databases

Relational Database Model

Edgar Frank Codd, a computer scientist employed by IBM in the 1970s, invented the relational database model in use today. The organization of a relational database is based on the relationship (i.e., context) of data into one or more tables (i.e., relations) of rows (i.e., tuples) and columns (i.e., attributes), with a primary key for each row.

Each entity type stored in a database has its own table, with the rows representing instances of that type of entity and the columns representing values attributed to that instance. Since each row in a table has its own unique key (i.e., primary key), rows in a table can be linked to rows in other tables by storing the rows' unique keys (foreign keys). Most all relational databases use structured query language (SQL) for querying, updating and retrieving the stored data.

The longevity and widespread use of relational databases is due to a number of factors, such as flexibility, simplicity, robustness, scalability, performance and compatibility with other applications in managing a large amount of data. However, in the field of data science and specifically data mining, although most relational databases store information in multiple databases, some algorithms operate on a single table extraction or relation.

A relatively simple SELECT statement (i.e., reading, retrieving) could have dozens of potential query execution paths that would be evaluated at run time. These operations are hidden to users—hence the simplicity—but under the hood, the relational database management system (RDBMS) determines the best execution path by using built-in logic like execution time cost-based algorithms.

When a database architect first designs a database, it is important to avoid common data pitfalls with any organizational initiative. Those pitfalls are:

- Duplication of data
- Logical and structural anomalies (RDBMS are sensitive to structural changes)
- Loss of data integrity

One such correctional practice is data normalization. Data normalization is a structured process for organizing data into tables to preserve the relationships among the data. It can include conversion of an actual range of values into a standard range of values to have a predefined list for selection (i.e., a closed list). A closed list is used when the data values are known. If the data values are unknown, an open list may be used, but that may result in data being deemed unreliable and lacking integrity. Normalization is the organization of database tables (relations) and columns (attributes) to ensure that their dependencies are properly enforced by database integrity constraints.

Figure 2.1—Example of Data That Is Not Normalized

Tenants Table

TenantID	TenantName	Leases			
		LeaseID	LeaseStart	LeaseEnd	Status
6	John Smith	25	2019-04-22	2019-10-22	I
		31	2019-10-23	2019-12-31	I
		36	2020-01-01	2020-12-31	A
7	Louis Corola	38	2020-06-01	2020-12-31	A
8	Adam Williams	33	2019-11-23	2020-05-31	I
		35	2020-06-01	2020-12-31	A

A normalized equivalent of **figure 2.1** might look like **figure 2.2**.

Figure 2.2—Example of Normalized Data

Tenants Table

TenantID	TenantName
6	John Smith
7	Louis Corola
8	Adam Williams

Lease Table (query is ordered by TenantID)

TenantID	LeaseID	LeaseStart	LeaseEnd	Status	NumOccu
8	33	2019-11-23	2020-05-31	I	1
8	35	2020-06-01	2020-12-31	A	1
7	38	2020-06-01	2020-12-31	A	4
6	25	2020-04-22	2020-04-22	I	3
6	31	2019-10-23	2019-12-31	I	3
6	36	2020-01-01	2020-12-31	A	3

Figure 2.3 is an example of a relational database.

Figure 2.3—Relational Database Management System Tables and Rows

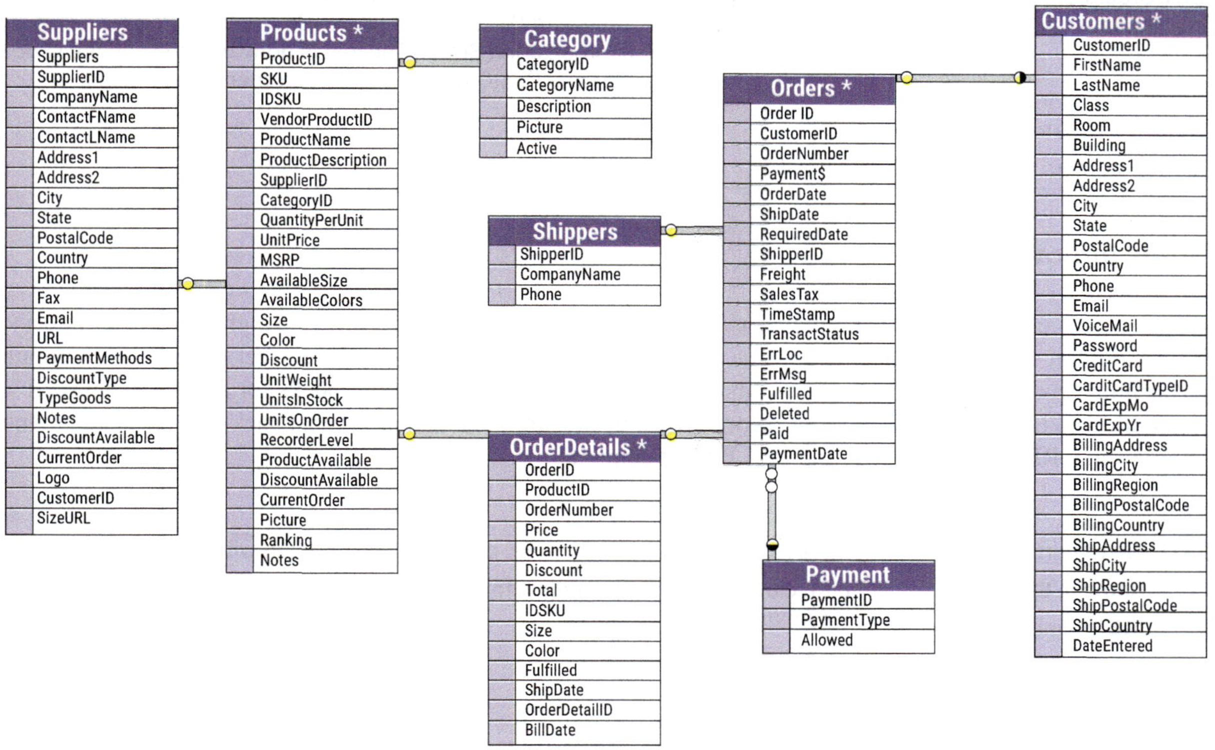

Source: Curtis, R.; "E-Commerce Database Design – Part I," Princeton University Extreme UltraDev, *www.princeton.edu/~rcurtis/ultradev/ecommdatabase.html*

A few basic structured query language (SQL) statements one should be familiar with are as follows:

Retrieving Information

A simple SELECT statement can be used as a wide net to get all columns from a single table:

SELECT * FROM Customers:	**SELECT** is a SQL keyword that signifies read-only search *(asterisk) is a wild card to signify all information from that table **FROM** is a SQL keyword that signifies the location (in this case the Customers table)

To retrieve very specific records, a SELECT statement can be combined with JOIN and WHERE statements where more than one table is used to obtain the records. There are four different types of JOIN:

- **(INNER) JOIN**—Returns records that have matching values in both tables.
- **LEFT (OUTER) JOIN**—Returns all records from the left table, and the matched records from the right table.
- **RIGHT (OUTER) JOIN**—Returns all records from the right table, and the matched records from the left table.
- **FULL (OUTER) JOIN**—Returns all records when there is a match in either left or right table.

Here is one example:

SELECT Orders,Orders.OrdID, Customers.CustName, Orders.OrdDate, Orders.OrdTotal FROM Orders INNER JOIN Customers ON Orders.OrdID=Customers.CustOrderID; WHERE Customers.CustName = 'John Smith';	In this query two tables are joined to return all the order ids (Orders.OrdID) and order totals (Orders.OrdTotal) for a specific customer: John Smith. The join is done on OrdID (in the Orders table) and CustOrderID (in the Customers table) fields.

Updating Information

Updating records can be done much quicker via a SQL statement than one at a time via a graphical user interface (GUI), although it can be extremely risky. The query should be vetted to ensure the correct records are manipulated by performing several SELECT statements first. Most often an application GUI checks for several business rules (e.g., not allowed to update OrderID post record creation) that are nonexistent in most back-end SQL statements. This applies to CREATE, UPDATE, and DELETE statements.

UPDATE Customers **SET CustName = 'Smithson'** **WHERE LIKE "%smi%";**	All records matching the WHERE clause will be updated with Smithson LIKE clause can be used with % to find all matching records such as 'alsmi', 'smith', 'pasmian', etc.

Deleting Information

Deleting records should be done sparingly, although applicable privacy laws and regulations should be considered as these rules may dictate retention requirements. If an audit trail is not captured, deleting could create orphaned records. Proper cleanup must be performed to delete the joining records, because a relational database stores the information in multiple tables. For example, deleting an order record may leave an active customer with a nonexistent order record, resulting in errors and possible GUI malfunctions.

DELETE FROM Customers WHERE CustID=6;	Deleting a specific record with Unique ID (UID) 6% can be used in the front or back of every string to expand or reduce the search result.

Creating Information

The following is an example of inserting a record into the Customers table without having to use an application GUI to do so.

INSERT INTO Customers (CustomerName, ContactName, Address, City, PostalCode, Country) VALUES ('Adam', 'Dean Williams', 'Jennifer Ct.', 'New Orleans', '77433', 'USA');	In parentheses are all the field names with corresponding input data, e.g., in CustomerName Adam will be inserted.

Non-Relational Database Model

A non-relational database, e.g., NoSQL, provides another option for data storage and retrieval. It is mostly used in big data applications.

NoSQL can easily capture all kinds of data, both structured and unstructured. It tolerates variations in new data types, and it is not disturbed by content structure changes. Unlike traditional databases, flat file document databases are not rigid, which means users can freely add fields to JSON documents without having to first define changes.

JSON, JavaScript Object Notation, is a simple document format that is readable by machines and humans alike; it is used to represent data in a schema-less format. A JSON file can be used to transfer information between completely separate computers and between programs residing on the same machine, offering superior information-sharing performance speed, flexibility, portability, and neutrality.

Due to the flat file structure, data is duplicated across different documents. For example, the personal data of a customer (first name, last name, billing address, etc.) would be duplicated in multiple order documents. Storage cost is a moot point, given that storage space is quite inexpensive.

Operations such as retrieving data (e.g., SELECT statements in SQL) are faster in NoSQL because they require no joining of tables to find order details—they are placed in one document.

Furthermore, NoSQL shares information across multiple clusters of computers without requiring application changes, unlike traditional SQL, which is limited to just one or a few database servers. Load balancing of mass datasets and queries is automatically distributed across servers, and when a server goes down, it can be quickly and transparently replaced with no application disruption. Also, a NoSQL database, such as MongoDB, offers an API for user-defined Map/Reduce methods. MapReduce is a parallel data programming framework designed for scalability and fault-tolerance, while an RDBMS, such as Oracle, does not support MapReduce.

Hadoop's ecosystem is commonly used by enterprises that have chosen to include the NoSQL model. Although Hadoop is continuing to evolve and may be replaced by other types of NoSQL models, this guide includes it as an example.

Hadoop, a software ecosystem used for data storage and computation, utilizes clusters of computers and NoSQL (see Chapter 1).[16] Hadoop includes the following software applications:

- **Apache Sqoop**—A command line tool for moving data from relational databases into Hadoop and back
- **Apache Flume**—Data collection software that can handle massive amounts of streaming data used for ingestion; robust and fault tolerant with tunable reliability mechanisms and many failover and recovery mechanisms
- **Apache Pig (also known as Pig Latin)**—Scripting language similar to SQL; used in conjunction with Hadoop instead of writing code in Java, can execute its Hadoop jobs in MapReduce, Apache Tez, or Apache Spark; used for data analysis tasks
- **Apache Spark**—A computational framework with an interface for programming entire clusters with implicit distributed data parallelism and fault tolerance; used for complex tasks such as machine learning
- **Apache Hive**—Data warehouse software that uses HiveQL; simplifies writing, managing and reading large datasets residing in distributed storage
- **Yarn (yet another resource negotiator)**—Used for resource management and job scheduling

There are two primary Hadoop modules:

1. **Hadoop distributed file system (HDFS)**—Distributes datasets across commodity hardware or clusters of low-cost and low-performance computers, sharing memory, computational power, storage and resources

[16] For a description of integrating Hadoop in an enterprise information system, see Banica, L.; A. Hagiu; "Big Data in Business Environment," *Scientific Bulletin - Economic Sciences*, Volume 14/Issue 1, *http://economic.upit.ro/repec/pdf/2015_1_10.pdf*.

2. **Hadoop mapReduce**—Distributes or delegates jobs to computers within clusters to perform work in parallel

MapReduce is the component Hadoop uses to distribute work around a cluster, so that operations can be run in parallel on different nodes. MapReduce is controlled by JobTracker software, which resides on the master node and monitors TaskTracker applications, running tasks at each node.[17]

The MapReduce system is based on two components:

- **Map**—Processes input key/value pair and produces a list of intermediate key/value pairs
- **Reduce**—Combines all intermediate values for a particular key and produces a set of merged output values (usually just one)

Figure 2.4 provides an example.

Figure 2.4—MongoDB MapReduce Example

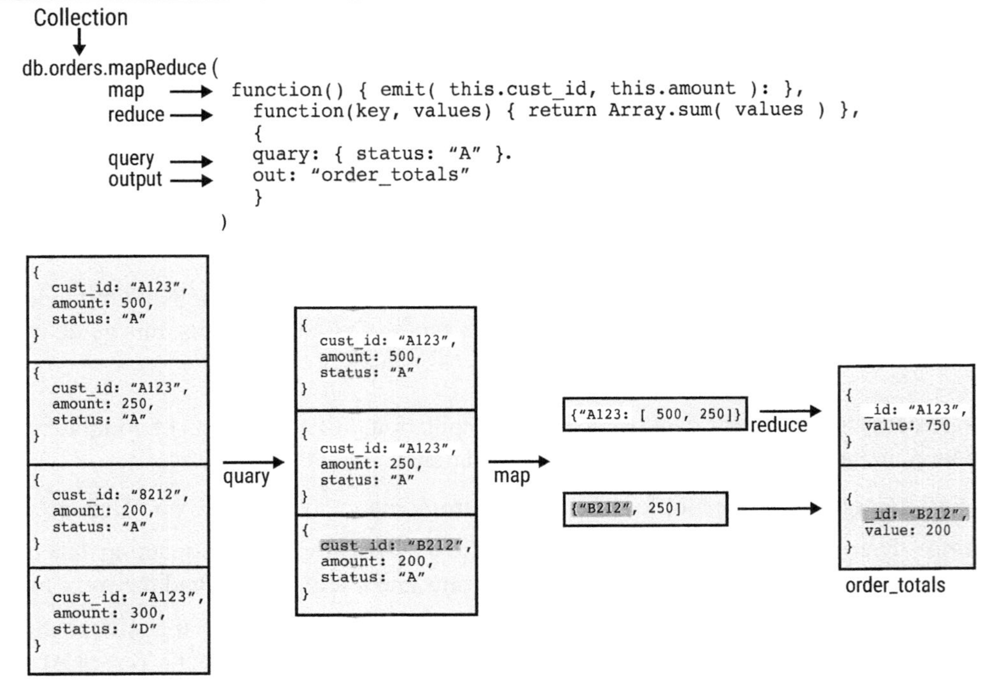

Source: Data Flair, "MongoDB Mapreduce Tutorial – Real-Time Example & Commands," 6 February 2019, *https://data-flair.training/blogs/mongodb-mapreduce/*

MapReduce is best for limited tasks like text searching or data mining, e.g., a search engine for unstructured data. MapReduce can scale to much larger data sets than SQL. In contrast, SQL's maturity and RDBMS structure allow good performance across a very broad range of complex relational operations, but it is lacking in scalability.

According to the CAP theorem, also known as Brewer's Theorem after its creator, Eric Brewer, a data store cannot provide more than two of the following benefits:

- **Consistency**—All clients are able to see and use the same data.
- **Availability**—The system is available and continues to show most recent data notwithstanding some failures.
- **Partition tolerance**—Despite network failures the system continues to operate.

[17] *Ibid.*

Figure 2.5 is a graphical representation of the CAP theorem.

Figure 2.5—The CAP Theorem

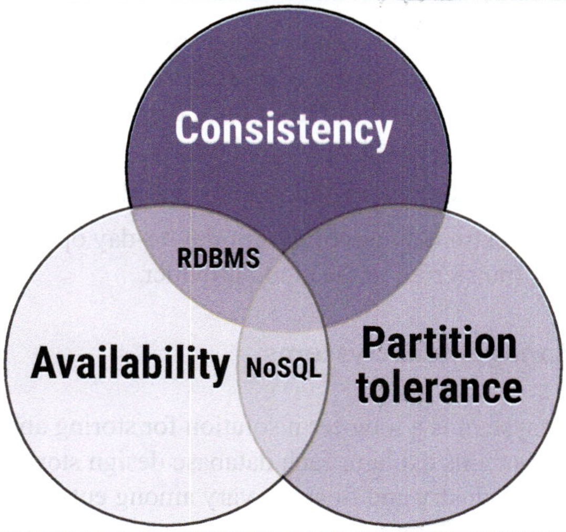

The differences between relational and non-relational databases are shown below in **figure 2.6**.

Figure 2.6—Differences Between NoSQL and SQL Database Management Systems

NoSQL (Nonrelational Database)	SQL (Relational Database)
Free and open source	Usually proprietary and requires expensive licenses
Duplication of data	No duplication
Flexible data structure	Rigid data structure
Fast	Slow
No downtime required to add/remove servers	Downtime required for server modification
Allows many concurrent users	Limits concurrent users
Can use clusters of inexpensive computers to share data and load balance	Usually requires expensive servers that are proprietary to the SQL provider
Supports data replication across clusters or data centers to ensure high availability and disaster recovery	Uses mirroring, which can be used for disaster recovery but requires additional setup and cost.
Lacks standardized interfaces and integrations	Mature and well-integrated; offers standardized interfaces
Command line interface (CLI)	Graphical user interface (GUI)
Will compromise consistency in lieu of availability	Will compromise availability in lieu of consistency

2.2.4 Autonomous Databases

In 2018 the first autonomous database (ADB) was launched by Oracle, a RDBMS company. ADB uses machine learning and is stored in a cloud to automate database security, backups, tuning, updates, and other routine management tasks traditionally performed by DBAs. Unlike a conventional database, an autonomous database performs all these tasks and more without human intervention, reducing human prone errors and costs for salaries and on-site infrastructures.

Autonomous transaction processing (ATP) can create, back up, patch and upgrade the database, while growing or shrinking the database as needed. It excels at handling large volumes of random data access and transactional workloads.

Autonomous database features include:

- **Security**—It will automatically download and install patches as soon as they are available to eliminate or reduce vulnerabilities.
- **Lower cost**—An advanced cloud infrastructure allows enterprises to forgo the high cost of database servers and their maintenance. In addition, database administrator or architect salaries can be eliminated, because the autonomous database is charged with all overhead tasks once performed by a DBA. Moreover, licensing is reduced and is converted to subscription-based cost based on use.
- **Experience**—A mature company knows how to optimize for speed, response and scalability.
- **Uptime**—Database availability is extremely important for day-to-day operations of an enterprise. Human error, which can result in downtime, is transferred to the cloud provider.

2.2.5 Common Database Management Systems

At its core, a database management system is a long-term solution for storing and retrieving massive amounts of data in different forms. As the following sections explain, each database design stores information differently. Because data types, capabilities, requirements, industry and finances vary among enterprises, there are many database options to consider.

Figure 2.7 shows an overview of commonly used database management systems.

Figure 2.7—Overview of Commonly Used Database Management System

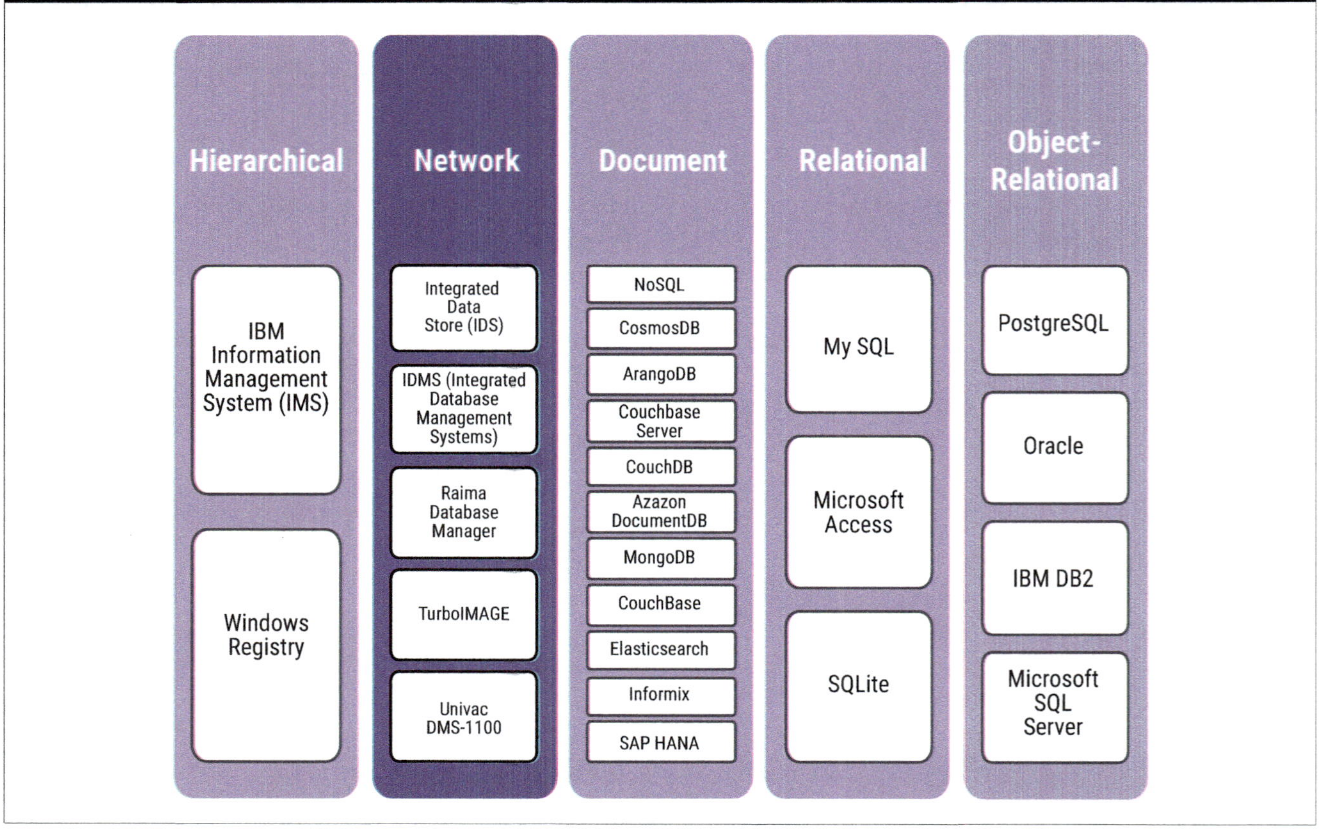

Choosing a Database Management System

Following are several factors an enterprise should consider when choosing a database management system. The priority of each depends on the industry and sector of the enterprise.

Deciding on the proper data model is dictated mostly by the data structure that will be stored. Is it structured, relational, easily categorized data—or not? Will this database be used for operational day-to-day transactional processing or mostly pure analysis? Does the response time matter? For decades, the relational concept was dominant, however document types, such as NoSQL databases, have gained recent traction in the marketplace.

- Relational databases require fixed schematics with a limited set of data types enforced for each row within a table. They require thorough modeling and design, and evaluation of the relationships between data and data attributes, assuming the database administrator is well aware of the data that will be stored. Otherwise, in some cases, a complete restructuring will be necessary.
- Hierarchical data structures store data in sequential files; document type databases provide greater speed of accessibility than relational databases.
- Document-oriented database management systems structure information with fields that may contain simple or complex values. Each document stored may have different fields, unlike SQL tables that require fixed schemas. As such, document-oriented database management systems offer extreme flexibility to developers, as fixed schemas do not need to be developed ahead of time. Some document databases use complex hierarchies of data with already-embedded documents.
- Object relational database management systems (ORDBMS) attempt to bridge the gap between relational and object-oriented modeling techniques in programming and extend the capabilities of relational databases. Borrowing from Java, C# and C++ they support complex data types, type inheritance and object behavior. In addition, while traditional RDBMSes focus on effective organization of data and limited data-types, ORDBMSes have a feature that allows developers to build and innovate their own data types and methods, which can be applied to the DBMS. This allows developers to view the problem area with increased abstraction.

Efficiency in the DBMS context equates with speed, or response time, for any CRUD activity a user or application may initiate.

- Document-oriented databases are optimized for data querying across one or more fields and support additional features, such as sorting by relevance and data faceting for drill-down capabilities. In a test performed in 2018, document-based MongoDB proved to be more efficient than Oracle DB in terms of average operation latency and throughput for each of the measured workloads.[18]
- Some databases are designed to optimize read-heavy apps, while others are designed to support write-heavy solutions. A database that can handle application input/output needs goes a long way toward enabling a scalable architecture.
- In some DBMSes, doing a read on a table row that is being updated will cause a lock or latency as the record is updated. Oracle and PostgreSQL are usually never blocked for read requests; other databases have improved significantly with later editions.

Databases may offer many capabilities and include every feature desired, but if they are difficult to use and manage they will likely stand idle. The user base for the database may include developers, network professionals, application integration vendors, and database administrators, all performing different functions. Every role needs easy-to-understand query language and functionality, and an intuitive UI in order to use the DBMS system efficiently.

- ORDBMS data are stored in a traditional database and manipulated and accessed using queries written in a query language like SQL. However, ORDBMS also supports object-oriented programming language, allowing object-oriented programming (OOP) developers to have a foundation for usability and integration.
- PL/SQL is a procedural language extension to SQL and a powerful way to build business rules in the database, although it is proprietary to IBM DB2. SQL/PL, another query language proprietary to Oracle, adds many procedural constructs to overcome some limitations of SQL. Microsoft SQL uses a completely different query language called Transact SQL (i.e., T-SQL). These language skills are not highly transferable, because each database has variations and a learning curve.

[18] Kolonko, K.; "Performance Comparison of the Most Popular Relational and Non-Relational Database Management Systems," Sweden, 2018

- Microsoft Access and SQL Server both offer an easy-to-use query designer, a drag-and-drop GUI, which is an intuitive and effective approach.
- XSQL, a combination of SQL and Extensible Markup Language (XML), is a database-independent language used for storing SQL queries, clauses and query results. XSQL can be used on Oracle and Microsoft databases.
- NoSQL databases offer different querying interfaces with no unifying standard.

Keeping data consistent becomes even more important as more sources of data flow into one centralized source of truth. Robust business rules that filter and rank data are a critical consideration, because the data source has to be trusted as factual and current for business processing, analytics and as an input into decision making. Atomicity, consistency, isolation and durability (ACID)—properties of a DBMS transaction set of statements that guarantee the validity of data—should be considered when choosing a database.

- **Atomicity**—To ensure validity, atomicity treats a transaction as a whole, whether it is multistep or not, before committing the record. For example, for a monetary transfer between accounts, one has to be credited and the other debited. Should either (credit or debit) fail, the transaction would not be committed to record.
- **Consistency**—Any CRUD operation performed on a data set must be within the constraints of the database rules, triggers, structures, etc., in order to ensure the data structure remains consistent and avoid database corruption. Using the monetary example above, consistency means that a historical log of the failed transaction would be created, and the two accounts would reflect their original balances.
- **Isolation**—Isolation ensures that concurrent CRUD requests to and from the database are carried out in the same manner as if they were performed sequentially to avoid botched or competing processes.
- **Durability**—Durability ensures that once a requested change is committed it will be persistent, even in the event of a failure, because nonvolatile memory is used.

NoSQL and other document-based databases choose availability over consistency, and most—if not all—are not ACID-compliant databases, but rather are **BASE**, i.e., basically available, soft state, eventual consistency databases.

The business security of internal data, customer data and vendor data is an essential aspect of any database implementation. It is especially important in industries that are held to a higher regulatory standard, including the government, health and financial sectors and to those enterprises that must comply with robust privacy rules.

Risks associated with data loss are not limited to electronic and digital threats such as hacking or data corruption (intentional malice or unintentional human error). They include physical storage risks, such as fire or theft, when the servers are housed locally.

Automatic data modeling and visualization and built-in robust reporting are advanced features in some DBMSes. NoSQL databases may not have the same rich functionality and lack reporting, which is considered a "nice to have" feature by some enterprises.

Many enterprises that began in the bricks-and-mortar age have operations, vendors and applications that are well integrated and work seamlessly with a RDBMS. Changing to a NoSQL database to take advantage of the greater processing power and reduced cost would involve many changes, such as rewriting of applications, which might not be feasible or cost-effective. There are many RDBMS integrations available off the shelf, but that is not the case with the document-based model.

As they are open source projects, several NoSQL databases can be used for free; many RDBMS alternatives are rather expensive by comparison. Additionally, faster application development and reduced development expenditures can be realized with less rigid data models with no explicit schema that requires more knowledge and expertise. Horizontal scalability of NoSQL is a major advantage. It enables the user to add nodes to a cluster dynamically without any downtime, resulting in linear effects on storage and overall processing capacities. There is no upper limit for the number of machines that can be added. RDBMSes were originally built to scale vertically, i.e., to add more hardware onto existing machines. However, vertical scaling has some limitations because it gives sublinear effects and sometimes requires high-risk operational efforts. There is an upper boundary that imposes restrictions on how much hardware can be added and efficiently utilized by the software.[19]

Frequency and automation of patches and software updates can affect uptime requirements. Databases such as IBM, Oracle, SAP, MS SQL Server, MS Access and others are more mature and sophisticated, and have the manpower to leverage support on a global scale. If the DBMS is in its infancy (document-based NoSQL) it may require constant updates that most likely will result in downtime and may need to be done manually after rigorous internal testing. MySQL, PostgreSQL and others are community-driven, free source databases, and support may be available through a forum or other online community support.

2.2.6 Data Lakes

A data lake is a centralized repository that stores data in a natural, raw state (**figure 2.8**). The data may be structured (e.g., rows and columns from a database input), semi-structured (e.g., XML, CSV, JSON) or completely unstructured (e.g., PDF, emails), or it may be in the form of blobs or files. A binary large object (BLOB) is a grouping of binary data treated and stored as a single entity in a database repository. Blobs are usually multimedia entities such as images, audio files or videos.

This large pool of unprocessed data receives input streams of data that flow from real-time internal and external sources, such as sales and CRM applications, inventory repositories, social media sources, operational databases, log files, clickstreams, Internet of Things (IoT) devices and others. Due to the lack of structure, massive storage is relatively inexpensive. The typical cloud-based repository uses parallel processing (i.e., computation for analysis) and data partitioning (i.e., storage). Those operations are usually kept separate so that they can be scaled independently, and users can pay for storage-only when they are not doing any analyses. Examples of cloud-based data lakes are Amazon, Google and Apache Hadoop, to name a few.

Typically, business knowledge of customers is limited to information an enterprise collects directly (e.g., through customers filling out forms when making purchases). The enterprise can obtain customer information indirectly by purchasing information from sources it usually would not have access to, and place it in a data lake for analyzing its customers in-depth.

Data scientists use data lakes for exploration and analytics, such as machine learning, and to provide opportunities for enterprises to learn about their customers by collecting their data from all sources. While in the data lake, data wrangling of the subsets can then be performed for ever-evolving purposes, implementing structure as it is analyzed and using temporary test areas to build and train models for any given task. Once processed, the data then move on to a resting database. The information derived from the data can be further utilized to make informed business decisions. Data wrangling (sometimes referred to as data munging) is the use of scripting languages, such as Perl, Python, and shell tools for data manipulation, formatting or restructuring to suit specific needs or increase decision-making power.

[19] Birgen, C.; "Advanced Process Simulation SQL vs. NoSQL," Norwegian University of Science and Technology, 2014

Figure 2.8—Data Lakes

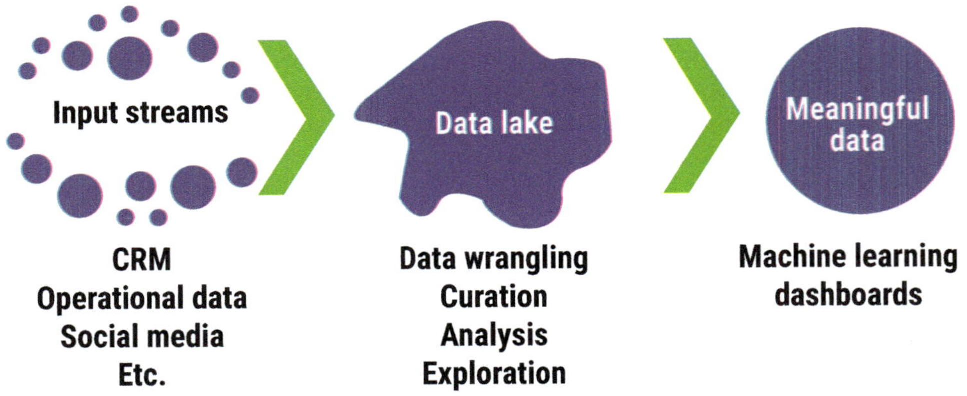

2.2.7 Data Warehouse

Data lakes and data warehouses are often used as synonyms, although they are not. The only real similarity between the two is their high-level purpose of storing data. Data lakes include raw data and both known and unknown data as input; data warehouses store scrubbed (i.e., cleaned and structured) data flowing mostly from internal data about the customer from multiple internal sources, such as financial, customer relationship management (CRM), pre- and post-sales, customer service calls, etc., for a specific use. This distinction is important, because data warehouses and data lakes serve different purposes and enterprises use them differently. While a data lake might suffice for one enterprise, a data warehouse would be a better fit for another. The one undeniable fact is that organizations that successfully generate business value from their data will outperform their peers.

Business intelligence (BI) platforms have long used data warehouses to store their data. BI platforms enable enterprises to build BI applications by providing several main capabilities: integrations to other platforms, online analytical processing (OLAP), built-in reports and dashboards, data tagging and metadata, and the ability to function as an information accessibility and delivery mechanism.

Typically, BI platforms allow general business users without deep data expertise to perform ad hoc analyses and reporting on structured data. Subsets of data go through the extract, transform and load (ETL) progression to establish business rules prior to being stored, so users do not have to impose structure on the data when they want to perform an analysis. The data are in the right form already and fit for purpose. Data are then made available continuously, along with analytics tools such as dashboards and preconfigured reports for specific usages, accessible via a user-friendly front-end GUI.

Data warehouses offer accuracy, timeliness and consistency of data, which usually are provided via traditional hardware and software, including column and row storage and RDBMS (**figure 2.10**). Often, data warehouse providers include service level agreements between IT and the enterprise to address maintenance, support response times or availability. A couple of competing BI reporting tools with powerful drag-and-drop visualizations are Tableau by Salesforce, and Power BI by Microsoft.

Figure 2.9 shows an overview of a data warehouse.

Figure 2.9—Data Warehouse

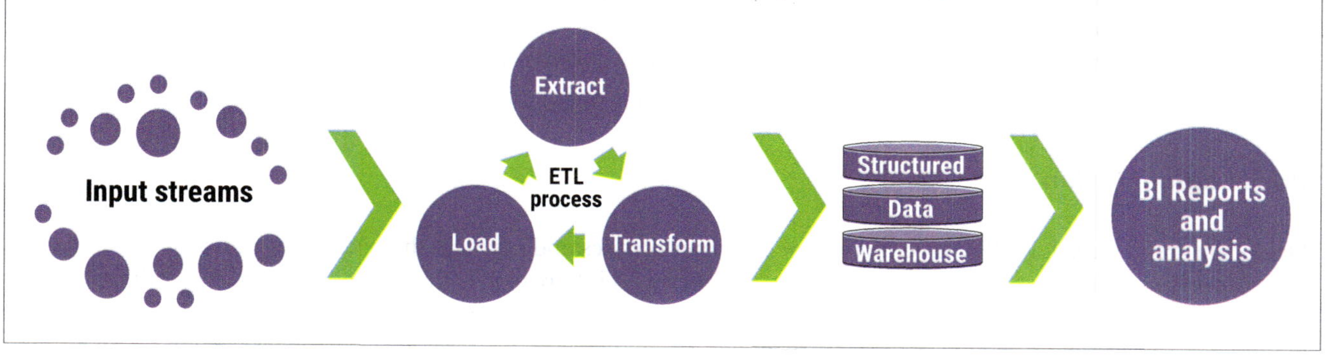

Figure 2.10 shows the main differences between data lakes and data warehouses.

Figure 2.10—Differences Between Data Lake and Data Warehouse

Characteristic	Data Lake	Data Warehouse
Data	Combinations of unstructured, structured and semi-structured	Structured, curated, schematic, "one source of truth"
Input sources	Internal and external, IoT devices, websites, social media, etc.	Internal, online transactional systems, operational databases and business applications
Data structure	Designed when curated and analyzed (schema on read)	Designed prior to data collection, and data flows into prestructured schema (schema on write)
Volume	Higher	Lower
Cost	Lower	Higher
Use cases	Exploratory, machine learning, predictive modeling, discovery and profiling	Dashboards, reports, ad hoc queries, preconfigured visualizations
Service level agreements (SLAs)	No	Yes
Tools used	Analytical modeling, workbenches, test sandboxes	BI reporting, batch reporting
Technology examples	Hadoop distributed file system, blog storage	RDBMS, batch reporting
Typical users	Data scientists and developers	Business analysts

2.2.8 Data Management Platforms

This section expands on two topics: data management platforms (DMPs), and customer data platforms (CDPs). Both of these are B2B (business-to-business) applications that enable enterprises to target marketing campaigns to subsets of existing customers and prospects.

Enterprises use data management platforms, or DMPs, for digital marketing and advertising to audiences that may be interested in their products and services. A DMP gathers data from a range of sources, both internal and external. Although DMPs pull from the internal data storage of business applications such as emails, CRMs, etc., the data are scrubbed and stripped of personal details, and consequently transformed into an anonymized entity. For external data, DMPs integrate with third-party brokers or corporate partners to obtain and share customer data.

Data collection, the first basic functionality of a DMP, involves a broad range of sources:

- **Offline internal sources**—CRM (customer relationship management) systems for pre- and post-sales data, financial systems or point-of-sale systems, customer service call logs, support tickets, etc.

- **Web and app sources**—Website traffic, mobile app data, events, conversions, digital cookies
- **Third-party data**—Data from other data collectors or vendors, including demographic or behavioral information via JavaScript applets, tracking tags or tracking pixels (e.g., Facebook Pixel)
- **Programmatic platforms**—Data on impressions served, conversions, video views from demand-side platforms (DSPs), and ad servers or other ad services (e.g. Google AdWords) the DMP used to serve an ad campaign
- **Marketing platforms**—Data on content that a visitor interacted with (e.g., marketing emails)

Data Segmentation

Once data are gathered, DMPs build individual customer profiles and segment them according to defined rules set by the enterprise's marketing team. The profiles are then grouped into personas that share attributes (e.g., all males over 40 years old who own a townhome and visited our handyman project site in the last two weeks; or all single females who live in Florida and filled out the free raffle at our Miami department store last December). Notice the specificity of the rules. These collections are consolidated into audiences, and all members of an audience (i.e., prospects) receive the marketing message that applies to them, resulting in a higher conversion rate. Conversion rate is the percentage of visiting website customers who actually purchase something. The formula is shown in **figure 2.11**.

Figure 2.11—Conversion Rate Formula

$$\text{Conversion rate (\%)} = \frac{\text{Total purchases}}{\text{Total visitors}} \text{ X } 100$$

There are two major ways a DMP can segment data:

- **Fixed segmentation**—This is usually constrained by a fixed date of an event, and a budget to produce one audience grouping. The criteria to define the audience are set by the user requesting the audience segment to target. For example, an airline may advertise reduced rates on a particular round trip from New York to California and may request the following audience segment:
  - All Internet users in New York and California
  - All New Yorkers who liked any fan page for any city in California
  - All Californians who liked any fan page for any city in New York

  In this example the data will be gathered from Facebook, Google website history and any other applicable third-party sites. This is ideal when an audience is behavioral, and historical data are not needed to make a selection.
- **Live segmentation (i.e., rolling segmentation)**—Live segmentation is a more involved process that is dependent upon the attribute's frequency and recency. A user request may look like this:
  - All Internet users in New York and California
  - All New Yorkers who liked any fan page for any city in California within the last 90 days *(recency)*
  - All Californians who liked any fan page for any city in New York within the last 90 days
  - All Californians and New Yorkers who entered three fan page comments within the last seven days *(recency and frequency)*

Data Activation

DMPs are nexus technologies—that is, they are conduits between data content repositories and content delivery ad sources. The third basic functionality of a DMP is that once live data is segmented, it is exported to a demand-side platform (DSP) or a remarketing platform. Ad space is purchased and launch of the advertising is scheduled. When the campaign is published, the user is given feedback and can optimize the request or requirements, which may increase or decrease the audience size in real time.

Since the data pushed into the platform is anonymous, most DMPs build segments and match data based on the probabilistic method. The probabilistic method is used because the identities are anonymous. The DMP uses an algorithmic assumption as to the identities' segmentation based on several data points.

There are three different categories of data:

- **Declared data**—Explicitly provided from users for an intended use (e.g., online forms or applications)
- **Observed data**—Digital footprints of users (i.e., search histories, geolocation information or types of web browsers used)
- **Inferred data**—Assumptions based on users' Internet behavior

A customer data platform (CDP) encompasses prepackaged software that creates a centralized database to store persistent customer data that marketers can leverage for targeted campaigns. The most distinct difference between a CDP and a DMP is that a CDP combines known (personally identifiable information, or PII, such as email, name and phone number) and unknown data into meaningful contextual data that is actionable without stripping any identifiable information, as a DMP does.

Moreover, a CDP creates a "story." It stitches together the life story of a person in digital form by creating nodes or contextual relationships between data to form a deterministic outcome of the individual. Using the deterministic method of data management enables enterprises to form accurate depictions of individuals, infer their attributes, attitudes, life cycle events, relationship to others and more. A CDP can also:

- Store ingested data indefinitely, subject to privacy restrictions (sometimes masking data)
- Ingest known and unknown data from any source (also extract context from web chat, email and phone data, and attach it to customer records)
- Create unified profiles of identified individuals
- Capture full details of ingested data
- Share data via integrations
- View individual customers multidimensionally in varying contexts (relationships, households, social networks, etc.) using visualizations
- Apply AI and machine learning to customer behavior

Figure 2.12 shows the main differences between a CDP and a DMP.

Figure 2.12—Differences Between Customer Data Platform and Data Management Platform

Characteristic	DMP	CDP
Data collection sources	Online and offline: • First party (direct from customer, enterprise internal CRM, or purchase transactions) • Second party (data from other enterprises) • Third-party data (multiple sources)	Same

Figure 2.12—Differences Between Customer Data Platform and Data Management Platform *(cont.)*

Characteristic	DMP	CDP
Intended target data	Cookies and segmented customer IDs from third-party providers	First-party data (structured, semi-structured and unstructured)
Privacy and anonymity	Yes	No
Data retention length	Limited by life of a cookie (30-90 days)	Indefinite
Storage of data	Two locations: DMP for data, and programmatic application for data activations and ads	One central database
Ease of integration	Lower	Higher
Targeting accuracy	Lower	Higher
Examples	Oracle's DMP (formerly known as BlueKai)	Informatica

Cloud-Based Big Data

Big data cloud services can manage volumes of both structured and unstructured data and offer increased processing power and storage a la carte. The cloud provides a readily available infrastructure, and the capability to scale fairly quickly to meet large spikes in traffic or usage. Each offering may include full transition of the enterprise's apps to the cloud, which may require architectural changes or a hybrid approach, with data science-related apps residing in the cloud and the rest kept on local servers.

The following are several of the big data cloud applications at the forefront of data analysis:

Google's Cloud Platform (GCP)

GCP is a suite of cloud computing services and applications that mirror the applications Google uses internally for its end-user products, such as Google Search, Gmail and YouTube. GCP uses the data from its web index to initially match queries with potentially useful results. This operation is augmented with data from trusted sources and other sites that have been ranked for accuracy by machine learning algorithms designed to assess the reliability of data.

Amazon Web Services (AWS)

At the center of Amazon's analytics offerings is AWS Elastic MapReduce (EMR), a managed Hadoop, Spark and Presto solution. EMR takes care of setting up an underlying Elastic Compute Cloud (EC2) cluster (virtual server) and provides integration with a number of AWS services, including Simple Storage Service (S3) and DynamoDB (key-value document database), which excel at ultra high-scaled operations similar to Amazon. EMR pricing is based on an hourly cost for each node plus the price of the underlying EC2 instance. To reduce costs, it is possible to take advantage of Amazon's EC2 reserved instance pricing to run jobs against a known schedule. Clusters can be created and deleted on demand to process specific jobs or kept running for extended periods of time. Clusters typically take about 15 minutes to provision before job execution begins, thus greatly reducing startup time compared to on-site setups.

Microsoft's Azure

Azure, by Microsoft, is an ecosystem of cloud-based applications such as Azure Data Warehouse, Azure Data Lake, Azure Data Factory, Spark, Power BI (used for dashboard reporting and visualizations), Azure Machine Learning, and Azure Synapse Analytics.

It offers a myriad of tools customized to fit the developer's skill sets. SQL (MPP—formally SQL Data Warehouse), or Spark SQL, and an SQL-like language are used to perform time series-based queries, and can call into Azure

Machine Learning to analyze data in real time. Open source tools such as HDInsight with Hadoop, Spark, Storm or HBase are also available, along with Microsoft's enterprise solution for building and running R models at scale.

Clusters can be automatically created and deleted on a schedule using PowerShell and automation. Alternatively, on-demand HDInsight clusters can be created for specific jobs invoked through Data Factory.

For processing real-time data, Azure has Stream Analytics. Stream Analytics can process data from BLOB storage or streamed through Event Hubs and IoT Hub.

Azure Data Factory exists as standalone service used to build data processing pipelines. Data Factory can read data from a range of Azure and third-party data sources, and through Data Management Gateway can connect and consume on-premises data. Data Factory comes with a range of activities that can run compute tasks in HDInsight, Azure Machine Learning, stored procedures, Data Lake and custom code running on Batch.

The platform has a standard and premium tier—the latter including the option of running Microsoft Machine Learning Server, Azure has a simple pricing model based on the number and type of nodes running. The node type governs the number of cores, RAM and disk space available on each node. HDInsight comes with a local HDFS and can connect to BLOB storage or Data Lake Store.

Pricing is based on the services used. For example, provisioned SQL is measured using its own performance measurement units, called Data Warehouse Units (DWUs), while the serverless offering is based on the amount of data that is processed per query.

2.3 Governance

To govern means to control or sway. Governance is a term used in many contexts. It can be broadly construed as enterprise governance, which ensures that enterprise objectives are achieved by evaluating stakeholder needs, conditions and options. It can mean setting direction through prioritization and decision making. It also can mean monitoring performance, compliance and progress against plans. In most enterprises, governance is the responsibility of the board of directors under the leadership of the chairperson.

In other contexts, governance refers to a more focused process, discipline or system. This section of the study guide focuses on data governance, which may at some points intercorrelate with enterprise or government-level governance.

2.3.1 Data Governance

Data governance is the management of the overall quality, integrity, relevance, privacy and security of available data. Like any custodian of information, the owner is required to protect the information and to balance the risk associated with possible data breaches or misuse against the value and profit anticipated from analyzing and using the data.

Data governance goals include:

- Manage all organizational data as enterprise assets
- Formalize accountabilities, and design roles and responsibilities
- Manage a scope that expands with organizational maturity

The first step in data governance is to choose a foundational framework. The framework should be flexible enough to adapt to the enterprise culture, suggest practical applications, include examples of successful implementations in different industries and offer scalability. Following are two such frameworks.

The Capability Maturity Model Integration (CMMI®) Institute, part of ISACA, published the Data Management Maturity (DMM)℠ model in 2014 to address the exponential growth of data by providing organizations with a solid framework already used in the field. The DMM model is based on the global CMMI model, pioneered by the US Department of Defense and used by thousands of enterprises and governments, with expanded focus on evaluating data within an enterprise.

The DMM is a comprehensive framework of data management practices in six key categories. It helps organizations benchmark their capabilities, identify strengths and gaps, and leverage their data assets to improve business performance.[20]

Figure 2.13 shows the six key areas of the Data Management Maturity model and their relationships.

Figure 2.13—CMMI Data Management Maturity Model

Source: CMMI Institute, "Data Management Maturity (DMM)℠ Model At-A-Glance," figure 1, page 4, *https://cmmiinstitute.com/getattachment/cb35800b-720f-4afe-93bf-86ccefb1fb17/attachment.aspx*

In addition to DMM by CMMI, ISACA's COBIT address governance at an enterprise level, which helps frame and navigate all subsequent governance models to align with three ultimate goals: benefits realization, risk optimization and resource optimization.

[20] CMMI Institute, "What is the Data Management Maturity (DMM) Program?," *https://cmmiinstitute.com/data-management-maturity*

The COBIT core model is shown in **figure 2.14**.

Figure 2.14—ISACA COBIT Core Model

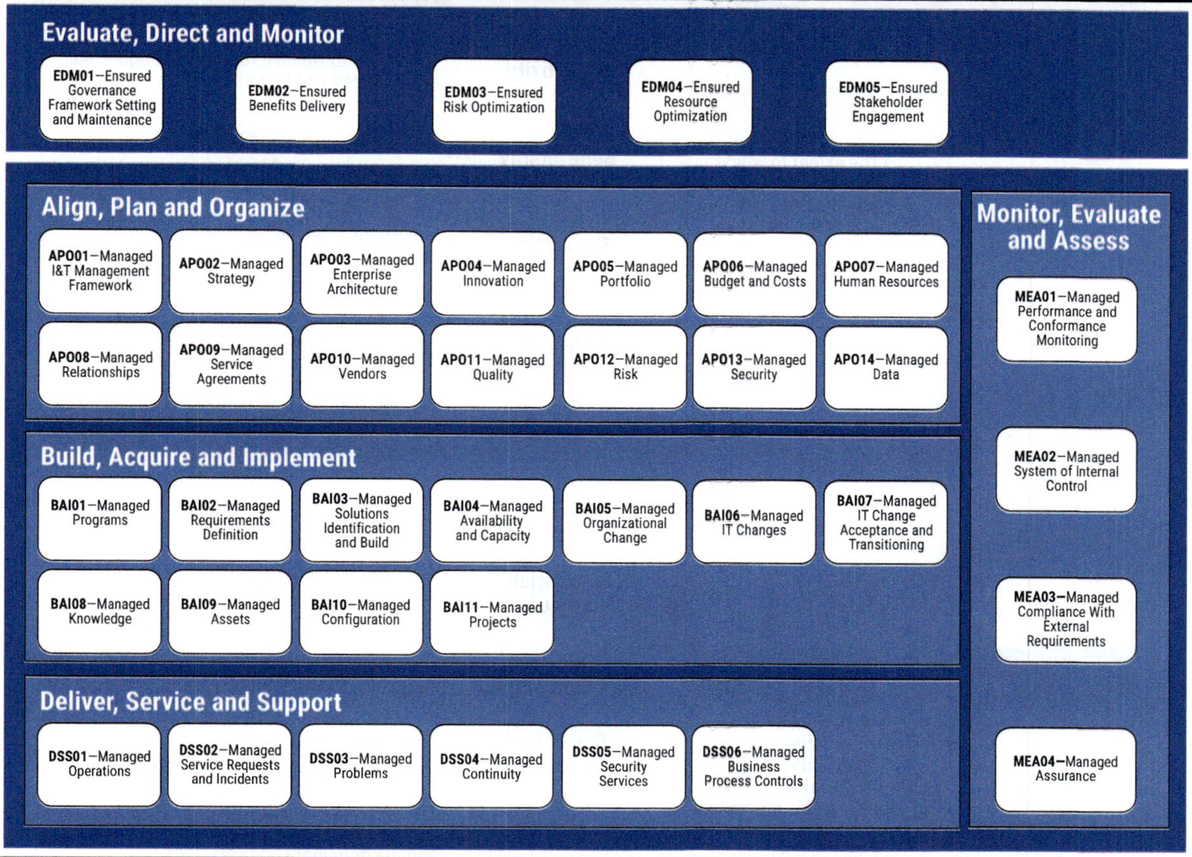

Source: ISACA, *COBIT 2019 Framework: Introduction and Methodology*, USA, 2018, figure 4.2, *https://www.isaca.org/bookstore/bookstore-cobit_19-digital/wcb19fim*

The frameworks mentioned can be used independently of each other or leveraged and customized for a holistic enterprise and data governance solution.

When considering data science within the greater scheme of the enterprise, it is necessary to consider the full hierarchy, roles and responsibilities, inhibitors and enablers, etc., required to adopt a well-balanced enterprise governance approach, with goal achievement supported by one of the previously mentioned frameworks. When viewing the data and information flow in isolation, it is important to consider quality criteria of the data, as shown in **figure 2.15**. Data is only valuable if the information can be relied upon.

Figure 2.15—ISACA COBIT Information Reference Model: Quality Criteria for Information

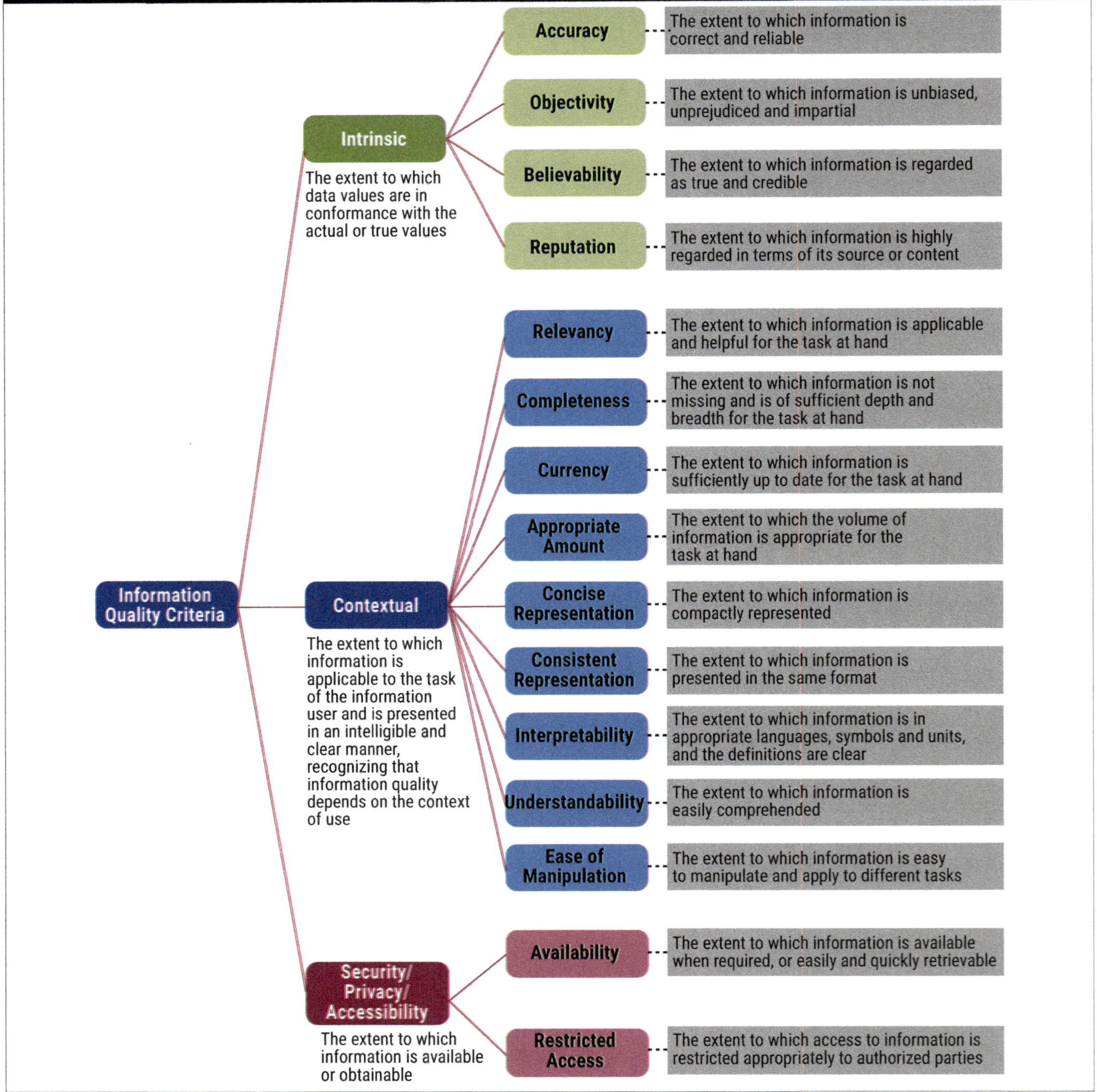

Source: ISACA, *COBIT 2019 Framework: Introduction and Methodology*, USA, 2018, figure 6.4, *https://www.isaca.org/bookstore/bookstore-cobit_19-digital/wcb19fim*

The goals of the information reference model are divided into three subdimensions of quality:

- **Intrinsic quality** references the extent to which data values are in conformance with the actual or true values. It includes:
  - **Accuracy**—The extent to which information is correct and reliable
  - **Objectivity**—The extent to which information is unbiased, unprejudiced and impartial
  - **Believability**—The extent to which information is regarded as true and credible
  - **Reputation**—The extent to which information is highly regarded in terms of its source or content

- **Contextual and representational quality** references the extent to which information is applicable to the task of the information user and is presented in an intelligible and clear manner, recognizing that information quality depends on the context of use. It includes:
  - **Relevancy**—The extent to which information is applicable and helpful for the task at hand
  - **Completeness**—The extent to which information is not missing and is of sufficient depth and breadth for the task at hand
  - **Currency**—The extent to which information is sufficiently up to date for the task at hand
  - **Appropriate amount of information**—The extent to which the volume of information is appropriate for the task at hand
  - **Concise representation**—The extent to which information is compactly represented
  - **Consistent representation**—The extent to which information is presented in the same format
  - **Interpretability**—The extent to which information is in appropriate languages, symbols and units, with clear definitions
  - **Understandability**—The extent to which information is easily comprehended
  - **Ease of manipulation**—The extent to which information is easy to manipulate and apply to different tasks
- **Security/privacy/accessibility quality** references the extent to which information is available or obtainable. It includes:
  - **Availability/timeliness**—The extent to which information is available when required, or easily and quickly retrievable
  - **Restricted access**—The extent to which access to information is restricted appropriately to authorized parties[21]

2.3.2 Legal and Regulatory Compliance

Digital tracking, as proliferated by CDPs, and to a lesser extent by DMPs, has raised many privacy concerns among institutions and individuals alike. More often than not, users create digital information for specific uses (e.g., applying for a job) not realizing their information may be used without their consent to sell them a product or a service, or to influence their emotions and eventually behaviors. Created in 2016, the General Data Protection Regulation 2016/679 (GDPR), which regulates data protection and privacy practices in the European Union, is meant to allow citizens to take control of their digital information. Privacy advocates fought and won when the European Union (EU) Court of Justice declared that collecting cookie information was a violation of fundamental rights of EU citizens. Every website within EU jurisdiction now must delete consumer data (upon request) and must inform users of the cookie capturing data on the site's landing page.

Regulations can walk a fine line, however, depending on how they are written. It can be a monumental task for individuals to request erasure of their data. An enterprise's right to process data may override an individual's "right to be forgotten" (as stated in Recitals 65 and 66 in the GDPR), and some organizations can dictate the time to comply (up to 30 days) and charge individuals a fee for the requested erasures.

In the United States, a similar law has been enacted in California. The California Consumer Privacy Act (CCPA), passed in 2018, grants consumers rights related to the collection, use and sale of their personal data—and it prevents businesses from discriminating against them for exercising those rights.

Under the CCPA, consumers have the right to:

- Ask about the categories and specific pieces of personal information a business has collected about them

[21] ISACA, *COBIT 5: Enabling Information*, USA, 2013

- Ask about the purposes for which the business uses that information
- Ask the business to delete personal information it has collected about them
- Request that their personal data not be sold to third parties[22]

California Civil Code section 1798.140(o) enumerates subtypes of personal information whose disclosure is prohibited, even via inference (this is not a comprehensive list).

> *(o)(1) "Personal information" means information that identifies, relates to, describes, is capable of being associated with, or could reasonably be linked, directly or indirectly, with a particular consumer or household. Personal information includes, but is not limited to, the following:*
>
> > *(A) Identifiers such as a real name, alias, postal address, unique personal identifier, online identifier Internet Protocol address, email address, account name, social security number, driver's license number, passport number, or other similar identifiers*
> > *(B) Customer records information, including name, signature, social security number, physical characteristics or description, address, telephone number, passport number, driver's license or state identification card number, insurance policy number, education, employment, employment history, bank account number, credit or debit card number, other financial information, medical information, health insurance information*
> > *(C) Characteristics of protected classifications under California or federal law*
> > *(D) Commercial information, including records of personal property, products or services purchased, obtained or considered, or other purchasing or consuming histories or tendencies*
> > *(E) Biometric information*
> > *(F) Internet or other electronic network activity information, including but not limited to browsing history, search history, and information regarding a consumer's interaction with an Internet website, application or advertisement*
> > *(G) Geolocation data*
> > *(H) Audio, electronic, visual, thermal, olfactory or similar information.*
> > *(I) Professional or employment-related information*
> > *(J) Education information, defined as information that is not publicly available personally identifiable information as defined in the Family Educational Rights and Privacy Act (20 U.S.C. section 1232g, 34 C.F.R. Part 99)*
> > *(K) Inferences drawn from any of the information identified in this subdivision to create a profile about a consumer reflecting the consumer's characteristics, psychological trends, preferences, predispositions, behavior, attitudes, intelligence, abilities and aptitudes.*
>
> *(2) "Personal information" does not include publicly available information. For purposes of this paragraph, "publicly available" means information that is lawfully made available from federal, state, or local government records. "Publicly available" does not mean biometric information collected by a business about a consumer without the consumer's knowledge.*
> *(3) "Personal information" does not include consumer information that is deidentified or aggregate consumer information.*[23]

Publicly available information, such as individuals' social media posts, and other public data, e.g., phone book listings, are not protected. Moreover, "publicly available" does not include consumer information that is anonymized via processing, or aggregate consumer information.

Following are other United States and global regulations that cover varying aspects of data protection for natural persons.

- **COPPA**—The Children's Online Privacy Protection Act enacted in 1998, which compels the FDA to protect minors aged 13 and younger online; in recent years YouTube, an online video platform, had to implement specific artificial intelligence (AI) algorithms that would flag videos that could be enticing to children to ensure the minor patrons/viewers:
  - Are minimally or not tracked via cookies
  - Are not targeted by advertisers

[22] Informatica, "CCPA Compliance," *www.informatica.com/solutions/data-governance-and-compliance/ccpa.html*
[23] Cal. Civ. Code §1798.140(o), *https://leginfo.legislature.ca.gov/faces/selectFromMultiples.xhtml?sectionNum=1798.140&lawCode=CIV*

- **SHIELD**—New York State's Stop Hacks and Improve Electronic Data Security Act
- **HIPAA**—The US Health Insurance Portability and Accountability Act passed in 1996
- **PISS**—Personal Information Security Specification, Singapore's Personal Data Protection Act
- **MIIT**—China's Ministry of Industry and Information Technology
- **PDPO**—Hong Kong's Personal Data Privacy Ordinance; shares some similar provisions with the EU's GDPR but includes the following glaring differences:
  - No requirement for parental consent for children under the age of 13
  - No right for natural persons to request data erasure or object to data profiling and processing

Unfortunately, existing regulations do not provide the intended protection of a natural person's digital data. Most consumers have to accept cookies to peruse websites and would have to retroactively opt out of any data collection, which is more cumbersome than simply opting in to desired marketing and data collection campaigns. For example, job seekers must check a box agreeing to long lists of uses of their data in order to apply. Hence individuals are forced to comply in order to conduct business, and are left without much choice. Moreover, these regulations are slanted toward DMPs, which collect temporary anonymous data that may reduce their value proposition, while CDPs, which collect first-party natural personal data indefinitely, continue to proliferate as regulatory standards evolve.

Figure 2.16 shows an example of the cookie tracking announcement on ISACA's home page. Each website must comply and include something similar. Not complying with this and other GDPR regulations can trigger fees of up to 20 million euro (~ US$23 million) or two percent of the enterprise's revenue for the preceding year.

Figure 2.16—Organizational Compliance With GDPR

Source: ISACA's home page with cookie notification at bottom.

Source: ISACA, *www.isaca.org*

2.3.3 Data Ethics

Most mature disciplines include a facet of ethics that is usually hard to ignore and must be defined for the discipline to mature and grow. Computer science and software engineering already incorporate a multitude of codes of conduct and ethical considerations that are well-defined. Though evolving, the field of big data ethics is still in its infancy.

Regulations, statutes and policies are mandated limitations, while ethics and codes of conduct are moral and ethical guidelines used to restrict enterprises' tendencies to use others' data for their own benefit and profit, without due consideration of the impacts of doing so.

Figure 2.17 includes a list of codes of conduct and ethics resources related to big data.

Figure 2.17—Big Data Code of Conduct and Ethics Resources

Organization	Resources
American Statistical Association	Ethical Guidelines for Statistical Practice (*www.amstat.org/ASA/Your-Career/Ethical-Guidelines-for-Statistical-Practice.aspx*)
Association for Computing Machinery (ACM)	ACM Code of Ethics and Professional Conduct (*www.acm.org/about-acm/acm-code-of-ethics-and-professional-conduct*)
British Computer Society	Code of Conduct for BCS Members (*www.bcs.org/upload/pdf/conduct.pdf*)
Data Science Association	Data Science Code of Professional Conduct (*www.datascienceassn.org/code-of-conduct.html*)
INFORMS for the Certified Analytics Professional	Code of Ethics for Certified Analytics Professionals (*www.informs.org/Sites/Certified-Analytics-Professional-Program/CAPs/CODE-OF-ETHICS*)
Institute of Electrical and Electronics Engineers (IEEE)	Ethics and Member Conduct (*https://www.ieee.org/about/compliance.html#code-of-ethics*)

Ethical considerations regarding software development applications for processing transactional day-to-day, operational data involve very limited data collection that is isolated, independent and has a very specific purpose—i.e., reason for collection. For example, customers are interested in electrical service at their new home.

It is expected that an electric company will ask customers for their full home address, contact information, and phone number to provide electrical service. Requiring a social security number to run a credit report as a qualification to receive service is also acceptable and logical for the intended purpose. However, if the electric company captured or asked for a list of appliances used and then sold that information to appliance manufacturing companies or warranty companies, that would be a violation of the intended purpose.

The focus on big data is concerned mainly with what is being processed, the nature of what is being processed, and who the processing is being done for or by. For example, big data has characteristics of volume, velocity and variety, which distinguish it from other information being processed, such as transaction data. In addition, big data projects could be for individuals, enterprises or clients. As a result, issues such as data confidentiality and privacy can be concerns.[24]

As with IT governance, the available frameworks offer guidance that can be adapted somewhat for big data, but that adaptation is likely to miss the specificity required, considering that data science involves individuals and their lives. Big data's owner (or perhaps *client*, according to the data scientist doing the work for the enterprise) can be interpreted many ways. It can be the enterprise actually collecting the data, the one aggregating it and selling it (e.g. CDPs), the third-party vendors that pass it from company to company via integrations, or the actual person who owns the data.

The regulations across states and countries add their own layers of complexity, because "personal" or "private" may have different sets of specifications associated with enterprise data management. An enterprise's own set of directives for compliance, and its own set of expectations for compliance reporting and auditing, add layers of conflict for the individuals and enterprises in this field.

Figure 2.18 depicts several rules from the comprehensive Data Science Association's Code of Conduct. It is a recommended resource for aspiring individuals in the data science domain in every role—especially data governance.

[24] O'Leary, D.E.; "Ethics for Big Data and Analytics," *IEEE Intelligent Systems*, vol. 31, no. 4, July-August 2016, DOI: 10.1109/MIS.2016.70

Figure 2.18—Sample Code of Conduct
1. Terminology
2. Competence
3. Abide by client decisions
4. No criminal or fraudulent activity
5. Communication with clients
6. Confidential information
7. Conflicts of interest
8. Duties to prospective clients
9. Data science evidence, quality of data, and quality of evidence
10. Misconduct
11. (Do not) engage in conduct that is prejudicial to methods of science
12. (Do not) misuse data science results to communicate a false reality or promote an illusion of understanding
Source: Adapted from the Data Science Association Code of Conduct Rules

In the enforcement of code of conduct rules, the data scientist is relied upon heavily to help the client (enterprise, employer, customer, vendor, etc.) understand the data's veracity, the algorithms' pros and cons, and the resulting conclusions. The data scientist cannot simply communicate haphazardly, but rather must exercise "reasonable diligence" when using words like "statistically significant," "causation" and "correlation," (Data Science Association Code of Conduct, rule 8.i) signifying the gravity of those words by stating a relationship. The words and any actions based on them can prompt an enterprise to make bad decisions. Following are a few pitfalls to consider:

- Data lacks veracity.
- Results are cherry-picked.
- Ethical inversion is employed; one's ethics are fitted to actions (or a profession) rather than the reverse.
- Replication of data science results fails; decisions are based on unreliable outcomes.
- Data is manipulated to meet a bias or desired result of the client.
- Results are misused to communicate a false reality or promote an illusion of understanding.[25]

2.3.4 Data Roles and Responsibilities

When considering how the data within an enterprise should be governed, it would be remiss to overlook the dedicated roles and responsibilities of the individuals across the enterprise who are accountable for enterprise data and affect the outcomes of its use. Varying business process definitions and misalignment across interdependent business units can adversely impact data quality, risk and security, and ultimately its value to the enterprise.

Borrowing from COBIT, the following **figure 2.19** reinforces the importance of the roles, activities and relationships between them as a foundational interrelated component.

[25] Data Science Association, "Code of Conduct, Data Science Code of Professional Conduct," *http://datascienceassn.org/code-of-conduct.html*

Figure 2.19—ISACA COBIT Governance Approach

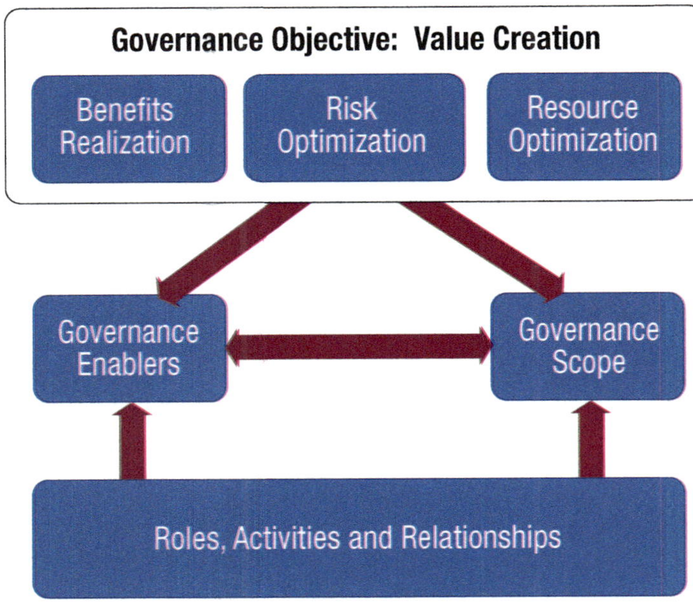

Source: ISACA, *COBIT® 5 (framework)*, USA, 2012, figure 8, *https://www.isaca.org/bookstore/cobit-5/wcb5*

Big data roles can be divided into three categories, aligning with the COBIT core model information and data objectives. **Figure 2.20** illustrates the roles and the assignment of governance objectives based on ISACA's COBIT core model utilizing some of the domains. When applying ISACA's COBIT core model to big data governance using governance and management objectives, the roles and tasks may differ across enterprises.

Figure 2.20—ISACA COBIT Governance Applied to Data Science Roles

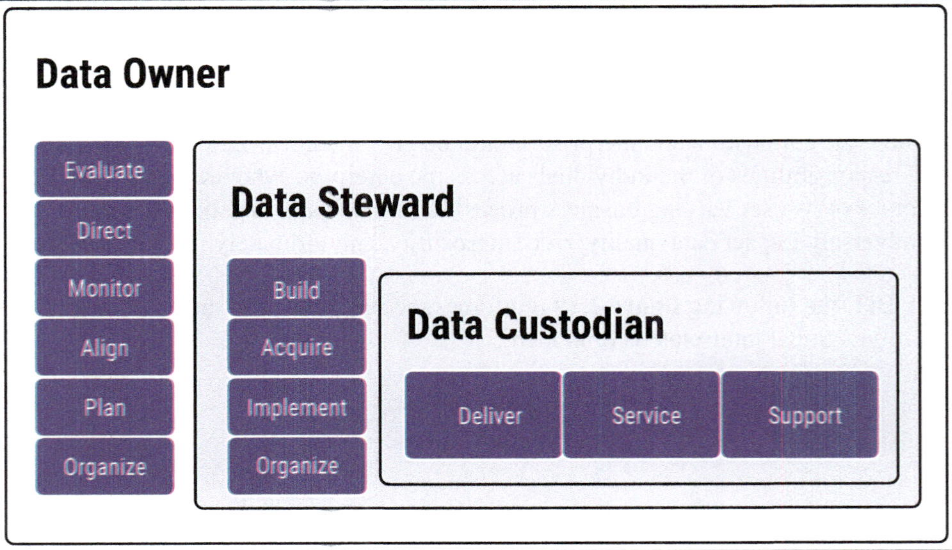

- Governance objectives—usually set by a board comprised of data owners, controllers and, to some degree, data processors—are grouped in the Evaluate, Direct and Monitor (EDM) domain under the direction of the board of directors and senior management. In this domain, the governing body and data owners evaluate strategic options, direct senior management on the chosen strategic options, and monitor the achievement of the information and data strategy. They may also monitor, evaluate and assess enterprise-level risk exposure from a compliance and regulatory perspective.

- Management objectives are grouped in four domains:
  - **Align, Plan and Organize (APO)**—Addresses the overall organization, strategy and supporting activities for information and technology (I&T)
  - **Build, Acquire and Implement (BAI)**—Treats the definition, acquisition and implementation of I&T solutions and their integration in business processes
  - **Deliver, Service and Support (DSS)**—Addresses the operational delivery and support of I&T services, including security
  - **Monitor, Evaluate and Assess (MEA)**—Addresses performance monitoring and conformance of I&T with internal performance targets, internal control objectives and external requirements[26]

Data Owner

A data owner is an individual who is accountable for a data asset—usually a director, manager or C-suite officer-level staff. There may be many data owners who form a data governance board (or several) for the enterprise. A sales director may own CRM data; be accountable for ensuring the data's quality, use and security; and be responsible for approving the data asset's flow to other systems. According to CMMI, data owners are also known as data trustees. Data trustees are accountable to the enterprise for security, privacy, data definitions, data quality, and compliance with data management policies and standards for a specific data domain.

Data Steward

A data steward's role essentially is to support the user community and assume responsibility for the day-to-day operations of the data assets. The user community includes both the producers of the data for the enterprise and the data consumers. The data steward is responsible for collecting and collating data, and evaluating related issues and problems. Typically, data steward assignments are based either on subject areas or line-of-business responsibilities. Data stewards are also known as business analysts, because they are usually the subject matter experts (SMEs) with respect to the business unit, its rules and its data. They can attest to the data's correctness from intrinsic and contextual perspectives, and enforce the data's business rules as outlined in **figure 2.20**. In lay terms, data stewards are responsible for *what* is stored in a data field.

Data Custodian

Data custodians are individuals and departments responsible for safeguarding computerized data. They are responsible for the safe custody, transport and storage of the data. They are tasked with the implementation of business rules. They maintain the technical environment and database structure, and they oversee the data's movement from one technological system to another. In **figure 2.20**, the data custodian role is responsible for the security, privacy and accessibility of the data and the systems that use the data assets.

Common job titles for data custodians are database administrator (DBA); data modeler; and extract, transform, load (ETL) developer.

[26] ISACA, *COBIT® 2019 Framework: Governance and Management Objectives*, USA, 2018, *www.isaca.org/bookstore/bookstore-cobit_19-digital/wcb19fgm*

Data Processor

A data processor, as defined in the European Union's GDPR Article 4, is a natural or legal person, public authority, agency or other body that processes personal data on behalf of a controller. A controller is defined as a natural or legal person, public authority, agency or other body that, alone or jointly with others, determines the purposes and means of the processing of personal data; where the purposes and means of such processing are determined by Union or Member State law, the controller or the specific criteria for its nomination may be provided for by Union or Member State law.[27]

In other words, a data processor can be any person or enterprise that does some data processing (reading, writing, manipulating, selling, storing, etc.). It really encompasses all the roles outlined above. That is to say, a database administrator—who is considered a data custodian—is also a data processor as defined by the GDPR. Although a DBA may be concerned with few aspects of data processing, the role is nonetheless considered a data processor performing tasks for the controller.

A few examples delineating data processor and controller both in task specificity and regulatory compliance accountability:

- An enterprise's marketing department processes and stores personal data of existing customers and prospects in its CRM application. It hires an online email marketing company to create email or ad campaigns (recall CDPs from section 2.2.8) that use the data. In this case, the marketing department is the controller and the email campaign company is the processor.
- An enterprise's HR department keeps records of all employees' personal information in an application and shares some of that data with the finance department of the organization. The HR department hires a third-party vendor to process the health benefits of the employees and shares the employee information to enable that service. In this example, the HR department is the controller and the third-party vendor is the processor.
- In order to carry out collections activities, an enterprise is provided with another enterprise's delinquent accounts' information. In this case the enterprise carrying out the collections is the processor and the enterprise that hired its services is the controller.

All the roles described may present with different titles or tasks at different enterprises, although the core responsibilities may align more often than not.

Data Scientist

The role of data scientist is often assigned to anyone who performs any activity that touches data, including data management, data processing, data analytics and so on. However, the skills required for these tasks vary greatly. As data science grows in usage, and teams grow in size, specialization naturally occurs. For example, data science teams often have people who focus on analytics (often called data scientists) and others who focus on collecting/cleaning data (often known as data engineers). In reality, many specializations are for "vertical" subject matter experts, such as data architects, big data engineers, data analysts or machine learning experts.[28]

According to the US National Institute of Standards and Technology (NIST), a data scientist is a practitioner who has sufficient knowledge in the overlapping regimes of business needs, domain knowledge, analytical skills, and software and systems engineering to manage the end-to-end data processes in the data life cycle.[29] The scientist has a high degree of technical skill and knowledge, usually with expertise in programming languages such as R and Python. Data scientists help businesses collect, compile, interpret, format, model, make predictions about, and

[27] GDPR, "Art. 4 GDPR Definitions," *https://gdpr-info.eu/art-4-gdpr/*
[28] Saltz, J.S.; Grady, N.W.; "The Ambiguity of Data Science Team Roles and the Need for a Data Science Workforce Framework," *2017 IEEE International Conference on Big Data*, USA, 2017, DOI: 10.1109/BigData.2017.8258190
[29] NIST, "NIST Big Data Interoperability Framework: Volume 1, Definitions"

manipulate all kinds of data in all manner of ways. They are experts at both construction and deconstruction.[30] A data scientist would need to be familiar with software development programming skills such as SAS, R and Python; application specifics such as Hadoop; SQL; and machine learning.

Figure 2.21 depicts the expected expertise for the three main roles in the data science domain.

Figure 2.21—Data Science Roles and Skills

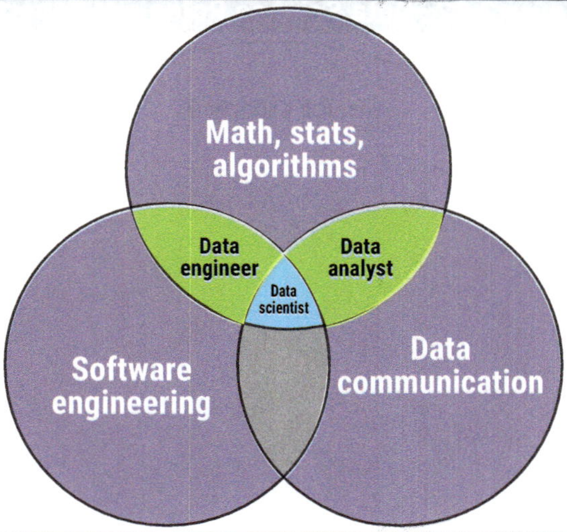

Source: Saltz, J.S.; N.W. Grady; "The Ambiguity of Data Science Team Roles and the Need for a Data Science Workforce Framework," *2017 IEEE International Conference on Big Data (Big Data)*, 11 December 2017, DOI: 10.1109/BigData.2017.8258190

Data Engineer

Data engineers manage exponentially growing stores of rapidly changing data and prepare data assets for use. They focus on data quality, development, deployment and management, and on optimization of data flows and infrastructure to transform and transfer assets to data scientists for querying. A data engineer needs to be familiar with programming languages such as Java, Scala, NoSQL, JSON, XML and Apache Hadoop, among many others.

Data Analyst

A data analyst is a data interpreter who typically specializes in identifying trends. Data analysts are similar to data scientists, without the coding experience. One way to think about data analysts is that they are junior data scientists on their way to becoming full-fledged data scientists. They provide questions that need answering and then organize and analyze data to find results that align with a high-level enterprise strategy. Data analysts are responsible for translating technical analysis to qualitative action items and effectively communicating their findings to diverse stakeholders. Data analysts bridge the gap between data scientists and business analysts. A few of the skills they may need are data wrangling, data mining and data visualization.

Business Analyst

A business analyst is very familiar with the business uses for data, and can be considered a data analyst in some contexts and enterprises. Business analysts recommend actions based on their experiences within an enterprise, or

[30] Eakins, A.; "30 Data Science Terms Explained in Plain English (with Examples)," *Springboard, www.springboard.com/blog/data-science-terms/*

with products and services, particular fields, or customers. They may be more articulate in describing the business impact of any decision to the final decision makers, members of the data governance board.

Figure 2.22 shows the expertise a well-known government defense contractor, SAIC, expected in the data science field. SAIC included "mission expertise," i.e., which points directly to business analysis- and data analysis-type skills related to knowing the mission—the business goals.

Figure 2.22—Data Science Roles and Expertise

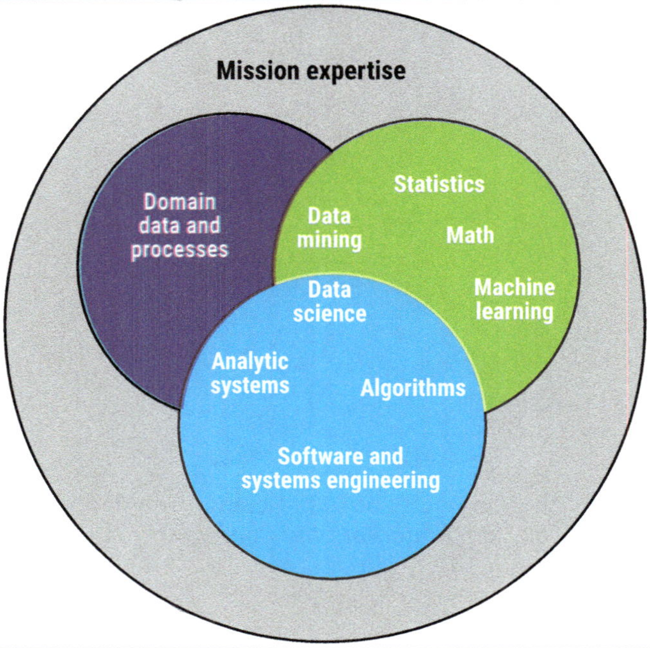

Source: Saltz, J.S.; N.W. Grady; "The Ambiguity of Data Science Team Roles and the Need for a Data Science Workforce Framework," *2017 IEEE International Conference on Big Data (Big Data)*, 11 December 2017, DOI: 10.1109/BigData.2017.8258190

When discussing the responsibilities associated with roles, the acronym RACI is used to denote responsible, accountable, consulted or informed for each task outlined in the project.

- **Responsible**—Refers to the person who must ensure that activities are completed successfully
- **Accountable**—Refers to the person who manages the work and is ultimately responsible for its completion
- **Consulted**—Refers to those people whose opinions are sought on an activity (two-way communication)
- **Informed**—Refers to those people who are kept up to date on the progress of an activity (one-way communication)

It is best practice to assign one responsibility per role, but in some cases a role may have more than one assignment.

Figure 2.23 is an example RACI chart, with specific tasks from the CMMI DMM model mapped to the COBIT framework. This chart serves as a guide—it does not include all the tasks and roles that would be used in a data science project.

Figure 2.23—Data Science Roles and Responsibilities RACI

Activity in DMM	COBIT Practice Area	Data Steward	Data Custodian	Business Analyst	Governance Board or Data Owner
AA 2.1 The target data architecture aligns with and complements the data management strategy.	APO03.02 Define reference architecture.	I	I/R	C	A
AA 2.4 A process is established and followed to ensure that data interface specifications are documented for shared data, with traceability from creation through consumption (end to end) by all sources within scope.	BAI08.02 Identify and classify sources of information.	R	C	I	A
PM 2.3 Data quality criteria are defined and embedded into service level agreements with both external and internal providers.	APO10.05 Monitor supplier performance and compliance.	I	R	C	A
PM 2.4 Planned discussions are held with data providers to address deviations to established data quality thresholds and targets defined in the service level agreement.	APO10.05 Monitor supplier performance and compliance.	I	R	C	A
MA 4.4 The performance of the data management attributes is analyzed and the data management baseline measures are maintained.	N/A	A	C	R	I

2.4 Access and Protection

There are layers upon layers of parallel and sequential processing in data science. Many input, output, entry and exit ports are used as data are transformed. Many organizations' goals include analytical and machine learning as central to processing data, yet not all organizations (or their vendors) address the associated security, accessibility and privacy issues surrounding data at each layer of transformation.

ISO/IEC 29100:2011 provides a privacy framework that specifies:

- A common privacy terminology
- The actors and their roles in processing personally identifiable information (PII)
- Privacy safeguarding considerations
- References to known privacy principles for information technology

ISO/IEC 29100:2011 is applicable to natural persons and organizations involved in specifying, procuring, architecting, designing, developing, testing, maintaining, administering and operating information and communication technology systems or services that require privacy controls for the processing of PII.[31]

[31] ISO, ISO/IEC 29100:2011 *Information technology – Security Techniques – Privacy Framework, www.iso.org/standard/45123.html*

- There are 11 substantive privacy principles:

1. Consent and choice
2. Purpose, legitimacy and specification
3. Collection limitation
4. Data minimization
5. Use, retention and disclosure limitation
6. Accuracy and quality
7. Openness, transparency and notice
8. Individual participation and access
9. Accountability
10. Information security
11. Privacy compliance

2.4.1 Data Accessibility and Protection

Big data schemes often encompass various components, applications, organizations, etc., in which a single security scheme has not been designed from the outset. Security and privacy measures for big data are increasingly stored on a public cloud infrastructure built by employing various hardware components, operating systems and analytical software. Traditional security approaches usually addressed small-scale systems holding static data on firewalled and semi-isolated networks. The surge in use of streaming cloud technology necessitates extremely rapid responses to security issues and threats.

Certain types of data thought to be too big for analysis, such as geospatial and video imaging, will become commodity big data sources. Those uses were not anticipated and adequate security and privacy measures may not have been implemented. Moreover, a huge increase in the number of sensor streams for the Internet of Things (e.g., smart medical devices, smart cities, smart homes) creates new vulnerabilities—in the Internet connectivity of the devices, in the data transport, and in the eventual data aggregation. This is one reason there has been the big push for 5G technology as it increases data throughput.

Most security and privacy methods were designed for batch or online transaction processing systems. Big data projects increasingly involve one or more streamed data sources used in conjunction with data at rest, creating unique security and privacy scenarios. Volatility is significant because big data scenarios assume the permanence of data by default. Security is a fast-moving field with multiple attack vectors and countermeasures. Data assets may be preserved beyond the lifetime of the security measures designed to protect them.[32]

Access control (AC) systems are among the most critical of network security components. It is more likely that privacy or security will be compromised due to the misconfiguration of access control policies than from a failure of a cryptographic primitive or protocol.[33]

Big data storage systems and processing are enabled via clusters of computers using open source applications such as NoSQL, etc., which were not as well organized or as considerate of security, accessibility and protection concerns as big enterprise implementations. In cluster data modeling there will likely be a Master System (MS) communicating with many Cooperated Systems (CS) in a trusted manner, passing data to and from each other. User authentication though authorization of specific jobs or data may be lacking.

[32] NIST, *NIST Big Data Interoperability Framework: Volume 4, Big Data Security and Privacy [Version 2]*, *https://www.nist.gov/publications/nist-big-data-interoperability-framework-volume-4-big-data-security-and-privacy-version*

[33] Hu, V.C.; T. Grance; D.F. Ferraiolo; D.R. Kuhn; "An Access Control Scheme for Big Data Processing," NIST, 1 January 2014, *https://csrc.nist.gov/csrc/media/projects/access-control-policy-and-implementation-guides/documents/big_data_control_access_7-10-2014.pdf*

Figure 2.24 shows a sample general cooperated model.

Figure 2.24—Cooperated System Model

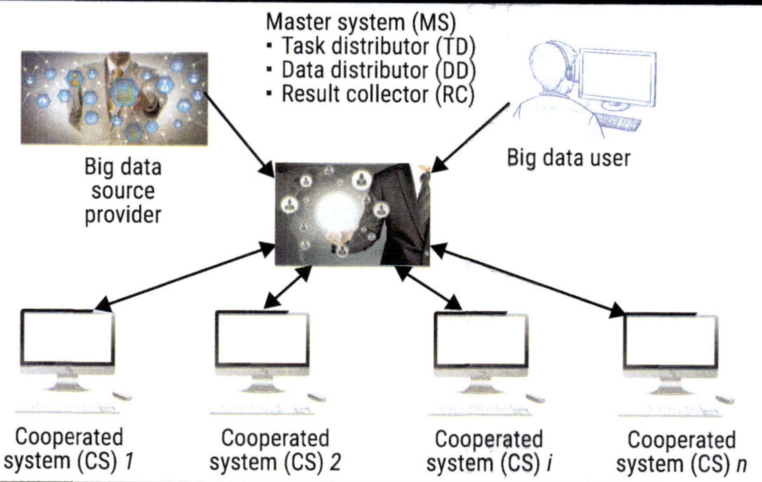

Source: Hu, V.C.; T. Grance; D.F. Ferraiolo; D.R. Kuhn; "An Access Control Scheme for Big Data Processing," National Institute of Standards and Technology (NIST), 1 January 2014, USA, *https://csrc.nist.gov/csrc/media/projects/access-control-policy-and-implementation-guides/documents/big_data_control_access_7-10-2014.pdf*

2.4.2 Managing Permissions

In any context, and especially in data science, there are layers of permissions that are determined based on need to know, purpose, authorization to the system, application, and finally data itself which must be compliant with the job or process and the user requesting it.

Permissions are multilayered. The permissions must be synchronized between the Master systems, data stores, third-party vendors and cooperated systems—a requirement that is quite complex and involved. What is more, it assumes the configuration to propagate permissions exists in all the external systems.

Figure 2.25 shows that in the case of Hadoop, each request may be broken into many smaller clusters of computational activities across many systems, exponentially multiplying the security and access control authentication propagation that would need to occur for each data node.

With this model, there are three main considerations that are critical for controlling access while keeping the value propositions of clusters intact:

- **Distributed computing**—Big data (BD) are processed anywhere resources are available, enabling massively parallel computation between MS and CSes. This creates complicated environments that are highly vulnerable to attack, as opposed to centralized repositories that are monolithic and easier to secure.
- **Fragmented/redundant data**—Data within BD clusters are fluid, with multiple copies moving to and from MS and CSes to ensure redundancy and resiliency. Data can become sliced into fragments that are shared across systems. This fragmentation adds complexity to data integrity and confidentiality.
- **Node-to-node communication**—MS and CSes usually communicate through unsecure protocols such as RPC over TCP/IP.[34]

[34] *Ibid.*

Figure 2.25—Distributed Cluster System Model (Hadoop)

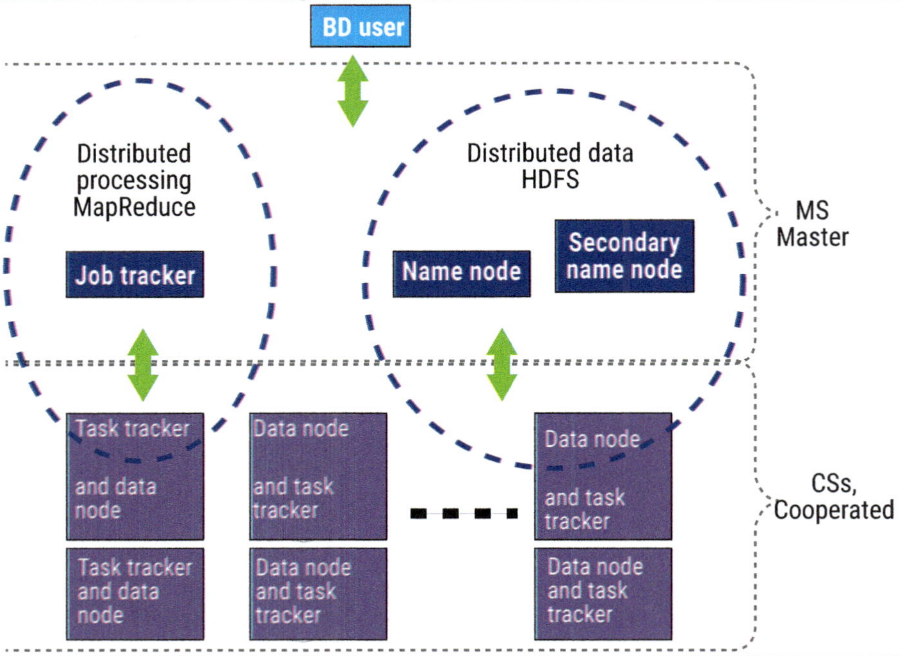

Source: Hu, V.C.; T. Grance; D.F. Ferraiolo; D.R. Kuhn; "An Access Control Scheme for Big Data Processing," National Institute of Standard and Technology (NIST), 1 January 2014, USA, DOI: 10.4108/icst.collaboratecom.2014.257649, *https://csrc.nist.gov/csrc/media/projects/access-control-policy-and-implementation-guides/documents/big_data_control_access_7-10-2014.pdf*

2.4.3 Third-Party and Vendor Access and Management

Security issues are multiplied when third-party vendors are involved. By definition, such vendors are outside the enterprise and likely have their own security protocols. In common practice, non-disclosure agreements (NDAs), also known as confidentiality agreements (CAs), do not necessarily cover electronic data storage breaches. As processors of data, vendors may have different requirements than the controller and owner of the data, so accountability may not be inherent. Reliance on periodic auditing, service level agreements (SLAs), and ironclad contracts that identify all risk transference and mitigations, should a data breach occur, may not be enough.

Additional consideration could include:

- **Data mismanagement**—Smaller firms are often subject to fewer regulations or lack mature governance practices. If not properly managed, they may degrade veracity because they are not primary information sources.
- **Data misinterpretations**—Inappropriate data interpretations may become more commonplace as researchers leverage third-party vendors that may not have access to the data, or to business analysts with the needed domain knowledge to interpret data contextually. Lack of common data schemas can further inhibit consistent security and privacy practices and lead to incorrect interpretations.
- **Transferability**—Consent for data usage may be restricted to the primary data owner and not transferable to another custodian, even one in the capacity of processor.
- **Global regulatory compliance**—As big data laws and regulations evolve, the admixture of controllers and processors can cross countries and states where the legalities for security and privacy compliance may differ, adding another dimension of complexity.

2.4.4 Data Obfuscation

Obfuscation is the deliberate act of creating source or machine code that is difficult for humans to understand. In the data science and personally identifiable information (PII) context, it means replacing true data with data that is inauthentic. For example, if someone's true birthdate is 5/5/75, obfuscation might replace that date with an automatically, randomly generated birthdate of 1/1/75, keeping the same data characteristic and complexity of the original birthdate.

According to the GDPR, obfuscation, also known as masking or pseudonymization, does not fulfill the regulatory compliance required for PII if the data can still be tied to a natural person by adding other information. "Personal data which have undergone pseudonymization, which could be attributed to a natural person by the use of additional information should be considered to be information on an identifiable natural person."[35]

Obfuscation has another meaning that individuals and organizations can use as a proactive measure: misinformation—i.e., using deception by supplying information that is outright false—as a means of protecting data. This is plausible when the correctness of the information is insignificant, such as in social media profiles, and it is a valid approach for consumers to retain their privacy. However, it has detrimental effects for data scientists who conduct analyses using bad data (GIGO).

The definition of privacy most widely used in the computer science community is derived from an 1890 paper, "The Right to Privacy," by Samuel Warren and Louis Brandeis. The authors referenced Judge Cooley, who called privacy the right "to be let alone."[36]

At the 2019 IEEE International Conference on Big Data, Shah Mahmood introduced the Anti-Data-Mining (ADM) Framework as a middle-ground option for both consumers and big data. Considering the universal data flow that is inherent to big data, Mahmood and the ADM offer a definition of privacy brings it into the 21st century:

> *Privacy is the right of an individual to control the flow of information, whether implicit or explicit, about themselves. An individual may temporarily lease their private data for specific purposes. However, any deviation from the purpose, or withholding private data beyond the lease term should be considered a breach of privacy.*[37]

The ADM introduced three different functions:

- **Sensitive data suppression**—Not including PII information such as date of birth, religion, etc.
- **Disinformation**—Generating misleading or false information in three ways: pseudo-information, location obfuscation and time distribution normalization to disrupt patterns (e.g., shopping online at all hours versus on weekends only)
- **Pseudo-information**—Fabricating information that looks real to the adversary; ideally, it should be statistically indistinguishable from real information

2.4.5 Tokenization

Tokenization is the process of replacing sensitive data with unique identification symbols that retain all the essential information without compromising data security.[38]

[35] GDPR, "Recital 26," *https://gdpr-info.eu/recitals/no-26/*

[36] Mahmood, S.; "The Anti-Data-Mining (ADM) Framework – Better Privacy on Online Social Networks and Beyond," *2019 IEEE International Conference on Big Data*, USA, 2019, DOI: 10.1109/BigData47090.2019.9006050

[37] *Ibid.*

[38] Lutkevich, B.; "Tokenization," *TechTarget, https://searchsecurity.techtarget.com/definition/tokenization?_ga=2.204940816.1440422422.1592690025-543538010.1592690025*

Replacing sensitive data can be done automatically using an algorithmic randomizer ensuring the same data type is used so algorithmic calculations are not affected by different data types. It is quite different from the encryption mechanism with ciphers changing the data type and rendering the information unreadable by some algorithms and database stores. Persistent (static) tokenization replaces authentic information with completely false information. Although it removes the identification relationship to the natural person, it also skews the data enough to skew big data analysis and inferences, crossing boundaries that exist in some algorithms.

Dynamic tokenization can be based on matching role, location or specific attributes to meet specific regulations that may differ according to PII subcategories or global location.

Tokenization is especially beneficial for:

- **Data type**—Structured data
- **Cost**—Lower total cost of ownership compared to encryption, because the tokenization process requires significantly fewer computational resources
- **Big data analysis**—Ability to draw conclusions from non-private information because only specific structured fields can be tokenized while the rest remain authentic and fully or partially visible
- **Key management**—None required, unlike encryption
- **Masking**—Using production data to populate test data, and replacing values

Tokenization can be used on simple structured data types, e.g., credit cards and social security numbers, and advanced types such as XML, JSON and others. Moreover, tokenization is useful for data in motion, such as cell tower communications.

2.4.6 Encryption

Encryption is the process of converting a plaintext message into a secure-coded form of text, called ciphertext. The ciphertext cannot be understood without converting it back to plaintext via decryption—the reverse process—which is done via a mathematical function and a special encryption/decryption password called the key. In many countries, encryption is subject to governmental laws and regulations that limit the key size or specify what may not be encrypted.

Encryption is part of a broader science of secret languages called cryptography, which is generally used to:

- Protect information stored on computers from unauthorized viewing and manipulation
- Protect data in transit over networks from unauthorized interception and manipulation
- Deter and detect accidental or intentional alterations of data, also known as Secure Hash Algorithms (SHAs)
- Verify authenticity of a transaction or document using SHA (160-bit hash), file check function or MD5 (128-bit hash)

Key elements of cryptographic systems include:

- **Encryption algorithm**—Mathematically based function or calculation that encrypts or decrypts data
- **Encryption key**—Piece of information similar to a password that makes the encryption or decryption process unique. A user needs the correct key to access or decipher a message, because the wrong key converts the message into an unreadable form.
- **Key length**—Predetermined key length. The longer the key, the more difficult it is to compromise in a brute force attack that attempts all possible key combinations.

Effective cryptographic systems depend upon a variety of factors including:

- Algorithm strength

- Secrecy and difficulty of compromising a key
- Nonexistence of back doors that allow encrypted files to be decrypted without accessing keys
- Inability to decrypt parts of a ciphertext message to prevent known-plaintext attacks
- Properties of the plaintext known by a perpetrator

There are two types of cryptographic systems:

- **Symmetric key systems**—Use single, secret, bidirectional keys that encrypt and decrypt; one example is Advanced Encryption Standard (AES)
- **Asymmetric key systems**—Use pairs of unidirectional, complementary keys that only encrypt or decrypt; typically one key is secret and the other publicly known[39]

Figure 2.26 displays the NIST Big Data Reference Architecture (NBDRA) Volume 4, Security and Privacy. It includes classification of cryptographic technologies in terms of their relation to NBDRA, the features they support and the data visibility they enforce.

Figure 2.26—NIST Cryptographic Technologies in Data Science

Technology	Data Provider	Application Provider	Feature	Disability
Homomorphic encryption	Encrypts data	Stores encrypted data	Capability to perform computations	Only at data provider
Functional encryption	Encrypts data	Stores encrypted data	Capability to perform computations	Result of allowed computations visible at application provider
Access control policy-based encryption	Encrypts data	Stores encrypted data	No capability to perform computations	Only for entities that have a secret key satisfying the access control policy
Secure multiparty computation	Plaintext data	Stores plaintext data	Collaborative computation among multiple application providers	Application Providers do not learn others' inputs. They only learn the jointly computed function.
Blockchain	Plaintext or encrypted data	Decentralized	Immutable decentralized database	Transaction logging in a decentralized, untrusted environment
Hardware primitives for secure computations	Encrypts data	Stores encrypted data	Capability to perform computations; verified execution	Controllable visibility at application provider

Source: National Institute of Standards and Technology, *NIST Big Data Interoperability Framework: Volume 4, Security and Privacy*, USA, October 2019, *https://nvlpubs.nist.gov/nistpubs/SpecialPublications/NIST.SP.1500-4r2.pdf*

Figure 2.27 is a graphical representation of the difference between the three.

[39] ISACA, *Cybersecurity Fundamentals Study Guide, 2nd Edition,* USA, 2017, *www.isaca.org/bookstore/csx-certificate-exam-resources/csxg2*

Figure 2.27—Differences Between Encryption, Tokenization, and Masking

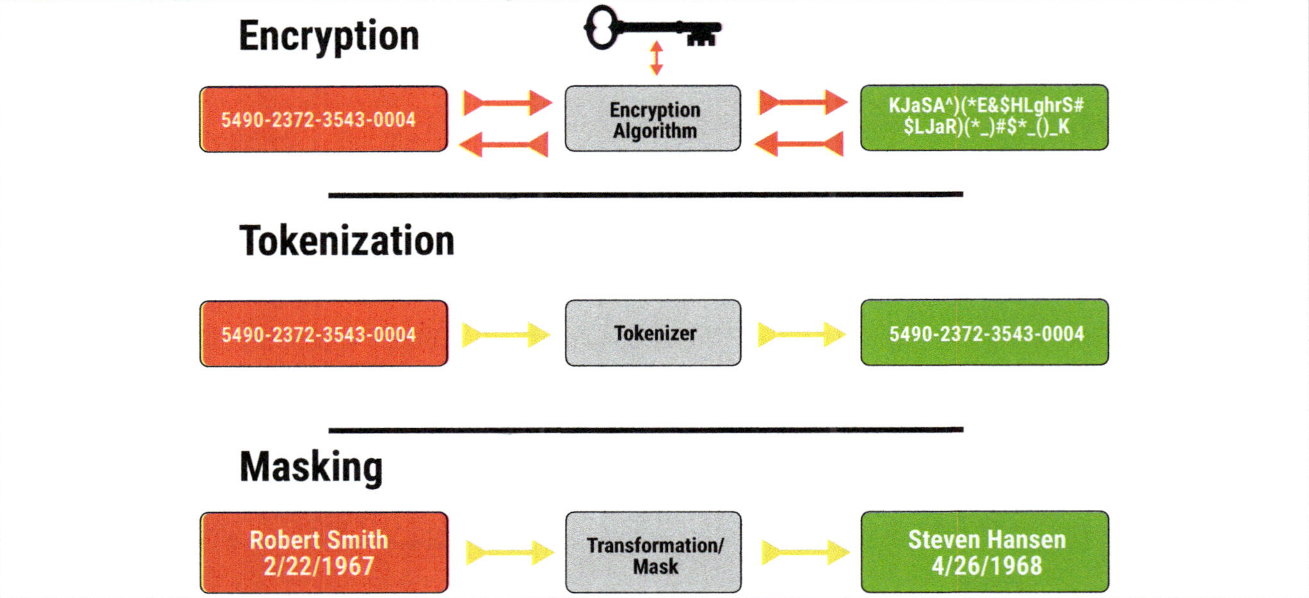

Source: Lane, A.; "Trends in Data Centric Security: Tools," *Securosis*, 9 July 2014, *https://securosis.com/blog/trends-in-data-centric-security-tools*

No data protection solution works for all data types. Specific use cases, enterprise risk tolerance, and applicable laws and regulations are all important factors when implementing a secure data framework without imposing restrictions that are prohibitive for proper analysis. In some cases, the breach of security occurs prior to entry into the data processing via data diddling. Data diddling is the practice of changing data with malicious intent before or during input into the system. Data leakages may occur at any place in the data flow from siphoning of the PII information by dumping computer files or stealing computer reports.

2.5 Chapter 2 Knowledge Check

Review Questions

1. Choose the **BEST** answer. One of the benefits of a NoSQL database is:

   A. Standardized interfaces
   B. Ability to use clusters of inexpensive computers to share data and load balance
   C. Graphical user interface (GUI)
   D. Ability to use mirroring, which can facilitate disaster recovery

2. Choose the **MOST** correct statement about data warehouses and data lakes:

   A. They both receive their data from the same sources.
   B. They are both data storage systems.
   C. Data lakes cannot receive unstructured data.
   D. Data warehouses are more cost effective.

3. Which of the following is characteristic of a data management platform (DMP)?

   A. Keeps data indefinitely
   B. Is more private than a CDP
   C. Has higher targeting accuracy
   D. Uses a centralized database

4. Choose the **BEST** answer. Data governance in data science is most important for:

   A. Ensuring data quality
   B. Implementing policies and procedures across the enterprise
   C. Assigning roles and responsibilities
   D. Access control

5. Protection of personally identifiable information (PII) can be accomplished by:

   A. Using distributed clusters instead of a centralized RDBMS
   B. Fragmenting the data
   C. Data obfuscation
   D. Applying the CAP theorem

Answers on page 88

Chapter 2 ANSWER KEY

Review Questions

1. A. Standardized interfaces
   B. Ability to use clusters of inexpensive computers to share data and load balance. Refer to page 55.
   C. Graphical user interface (GUI)
   D. Ability to use mirroring, which can facilitate disaster recovery

2. A. They both receive their data from the same sources.
   B. They are both data storage systems. Refer to page 60.
   C. Data lakes cannot receive unstructured data.
   D. Data warehouses are more cost effective.

3. A. Keeps data indefinitely
   B. More private than a CDP. Refer to page 64.
   C. Has higher targeting accuracy
   D. Uses a centralized database

4. **A. Ensuring data quality. Refer to pages 65-66.**
   B. Implementing policies and procedures across the enterprise
   C. Assigning roles and responsibilities
   D. Access control

5. A. Using distributed clusters instead of a centralized RDBMS
   B. Fragmenting the data
   C. Data obfuscation. Refer to page 83.
   D. Applying the CAP theorem

DATA SCIENCE
FUNDAMENTALS
An ISACA Certificate

Chapter 3:

Data Science Process

Data Science Process

Data Science Process

3.1 Learning Objectives

After completing this chapter, learners will be able to:

1. Identify open and cross-industry standards used to process data.
2. Describe techniques used to collect data.
3. Explain activities performed to prepare data for analysis, categorization and modeling.
4. Identify methods to uncover relationships among data.
5. Identify tools used to build, model and analyze data.
6. Describe concepts related to business analytics.
7. Distinguish among types of machine learning algorithms.
8. Distinguish among types of visualization and reporting tools.

3.2 Data Discovery and Goal Identification

There are 2.5 quintillion bytes of data created every day, with the last 90 percent created in the last two years.[40]

These staggering numbers could cause some to become overwhelmed—not sure what to collect, where to capture the information, how to process it, or why to collect some related and seemingly unrelated data.

One thing is known: The enterprise's ultimate goal is to produce valuable insights in order to sell more products or services.

Five companies in 1996 formed a joint effort that resulted in what is now known as the cross-industry standard process for data mining (CRISP-DM) open standard process. The cross-industry standard process for data mining included input from SPSS (an AI development company), Teradata (a database and analytics company), Daimler (aka Mercedes, a multinational automotive corporation), NCR (national cash register software company) and OHRA (an insurance company). The European Union's European Strategic Program on Research in Information Technology (ESPRIT) sponsored the effort.

The consortium, which included representatives from global enterprises in a myriad of industries interested in data mining, shared experiences regarding massive amounts of data.

The CRISP-DM model has six phases:

- Business understanding
- Data understanding
- Data preparation
- Modeling
- Evaluation
- Deployment

[40] Marr, B.; "How Much Data Do We Create Every Day? The Mind-Blowing Stats Everyone Should Read," *Forbes*, *www.forbes.com/sites/bernardmarr/2018/05/21/how-much-data-do-we-create-every-day-the-mind-blowing-stats-everyone-should-read/#de8e55160ba9*

Figure 3.1 depicts the original CRISP-DM model.

Figure 3.1—CRISP-DM Model

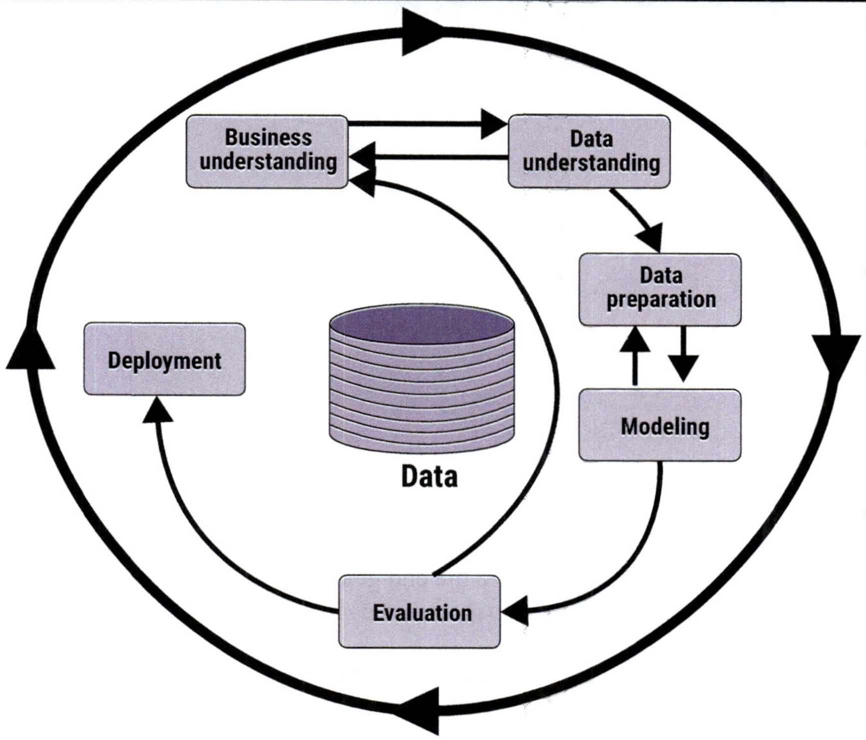

Source: Smart Vision Europe, "An overview of the CRISP DM methodology," *https://smartvision-me.com/the-crisp-dm-data-mining-methodology/*

The CRISP-DM model was refined and expanded upon by IBM with the release of Analytics Solutions Unified Method for Data Mining/Predictive Analytics (ASUM-DM).

3.2.1 Requirements and Resources

The first step of every project plan is understanding the project's purpose for existing, also known as the project's desired end goal or the business case. Data science projects are no different in that respect. The CRISP-DM subtasks, resources, tools and role-naming conventions may differ, but the fundamental process incorporates logical, common sensical steps, as seen in related disciplines (**figure 3.2**).

Figure 3.2—Project Phases: A Comparison

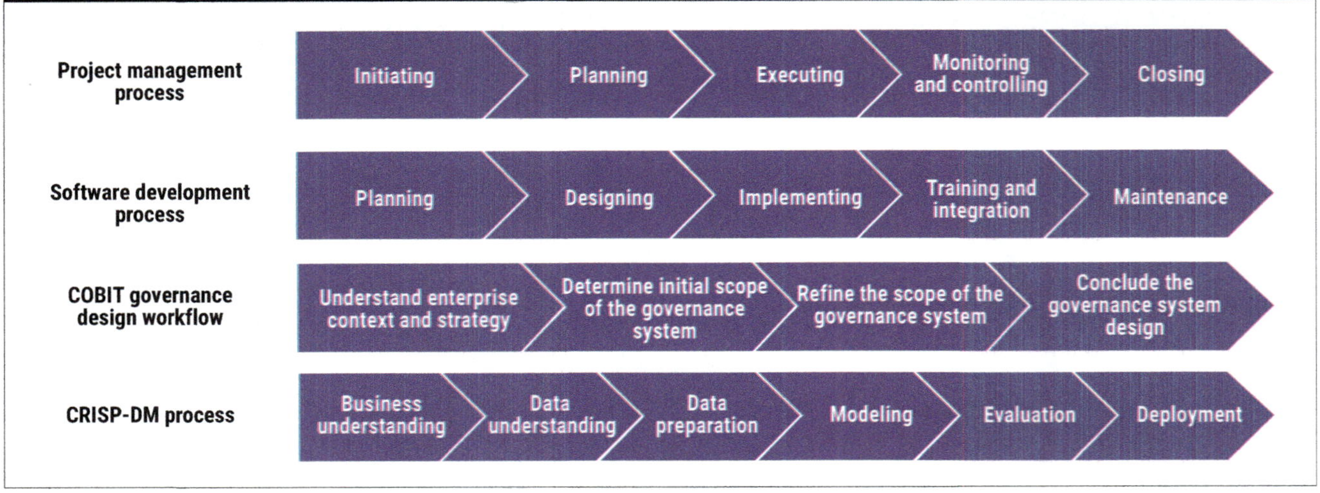

Business understanding is the most critical phase of the CRISP-DM model. Projects often fail due to very vague and abstract goals, because requirements have not been explicitly defined. This step can be understood as the interrogation phase. Asking lots of questions of stakeholders can help them further define goals to identify granular, measurable, concrete requirements, and help optimize the end goal. This phase includes subject matter experts such as business analysts, sales professionals, project managers and data analysts to compile the information.

Following is a list of tasks that should be considered during the business understanding phase. It is not meant to be exhaustive, because industry, enterprise and project specifics will differ.

- Thoroughly understand the problem to be solved and questions to be answered.
- Determine organizational objectives, products and services, and define the KPI criteria of the project and how performance will be measured and gauged. Moreover, identify the risk tolerance of the department and stakeholders sponsoring the big data project. Learn about the enterprise's industry and regulations background at least enough to ask when the legal department should be consulted.
- Assess the enterprise's culture and project sponsors fully, inventory the resources available, including personnel skills, and find out whether there will be dedicated resources for the project.
- Define data requirements, and ensure that a requirements analysis follows best practices for writing testable requirements (see **figure 3.3**) that trace back to enterprise and project goals for value realization.
- Define data mining goals, measurement tools, data resources and success criteria.
- Create a project plan outlining resources assigned, sponsors, planned communication plan, planned goals, cost and benefits analysis of the project's viability, risk and terminology. The project plan should be a living document, growing and changing as the project progresses, allowing for flexibility and the ability to pivot if the data uncovered should change the requirements or expected risk.

Figure 3.3—Requirement Characteristics

Requirement Characteristics	Explanation
Clear (concise, simple, defined) and understandable	Requirements should not contain unnecessary verbiage or information.
Unambiguous	There should be only one way to interpret the requirement. All acronyms should be spelled out.
Testable (verifiable)	During the testing and integration phase, testers should be able to verify whether the requirement is implemented correctly. The test should either pass or fail. It is a good practice to not leave any term or sentence open to interpretation.
Correct	If a requirement contains facts, the facts should be true all the time; otherwise, each should be a separate requirement.
Necessary and nonredundant	A requirement is not necessary if it is not needed by any stakeholders or if removing the requirement will not cause loss of functionality. Moreover, the same feature request (e.g., a pop-up calendar) should not be entered multiple times in the requirements document or user stories; better to detail the location of the calendar in one requirement.
Privacy benchmarks/ boundaries	Privacy, regulations and policies assessment may need to be applied to each requirement to assess any trade-offs (e.g., would you rather have more false negatives or more false positives?).

Source: Adapted from ISACA, *Software Development Fundamentals Study Guide*, USA, 2021, *https://www.isaca.org/bookstore/bookstore-misc-ebook/witassd*

3.2.2 Formulation of Hypotheses

A hypothesis is a supposition or proposed explanation made on the basis of limited evidence as a starting point pending further investigation. In other words, theories are based on observations. To know whether a theory is true or false it is necessary to ask questions and then devise tests. Compelling evidence will support a theory, while weak or no evidence will justify rejecting it.

A possible consequence of neglecting this step is to expend a great deal of effort producing the right answers to the wrong questions. There are several different types of hypotheses:

- **Simple hypothesis**—This hypothesis may be limited to personal experience, or word of mouth among a very limited set of the population the individual encounters. It is not based on data analysis. For example, earning a PhD leads to higher income.
- **Complex hypothesis**—This hypothesis scrutinizes the relationships between two or more independent and two or more dependent variables. For example, weather patterns and school schedules are responsible for traffic congestion, resulting in a higher number of accidents.
- **Empirical hypothesis**—This hypothesis relies on evidence that is produced using the scientific method, including observation and experimentation. For example, when there is a logfile, something executed.
- **Null hypothesis (notated by H0)**—This is a statement that shows no relationship between the variables, or indicates there is not enough information to form a question or statement. An alternative hypothesis will need to be created to disprove or provide knowledge to H0. Alternative hypothesis is denoted with H_1 or H_a.

Forming a good hypothesis (i.e., expressing the problem via an assumptive statement or question that needs answering) requires that individuals:

1. Ask lots of questions to form a clear *logical* statement of the problem, or brainstorm the possible variables that make up the problem and continuously break down the question to granular details. *If the final statement cannot be falsified or confirmed by gathering data for testing, it is not a hypothesis.*

2. Test the hypothesis using one of the following analytical methods (the chosen approach determines the data collection contents and formats):
   a. **Regression**—Is it possible to predict the sales revenue of product A for next year?
   b. **Classification**—Do customers prefer Product A, B or C?
   c. **Clustering**—Do Products A and B perform better in India, Pakistan and Germany than in the United States?
3. Identify outliers and produce more questions to learn why those outliers exist.
4. Choose the techniques that will be used to analyze the data needed to test the hypothesis, whether statistical or machine learning:
   a. Linear regression
   b. Logistic regression
   c. Clustering
      i. K-means
      ii. Hierarchical
      iii. Density based
   d. Classification trees
   e. Random forests
   f. Neural networks
   g. Text mining
   h. Principal component analysis
   i. Support Vector Machines
   j. Hidden Markov Models

   Experiment by testing to create data, or explore existing data to prove or disprove the claim using the techniques outlined in step 4.

3.3 Data Collection

Based on a 2016 survey, data scientists typically spend about 80 percent of their time collecting and cleaning data in preparation for conducting an analysis, as shown in **figure 3.4**.

Figure 3.4—Survey of Data Scientist Time Allocation to Tasks

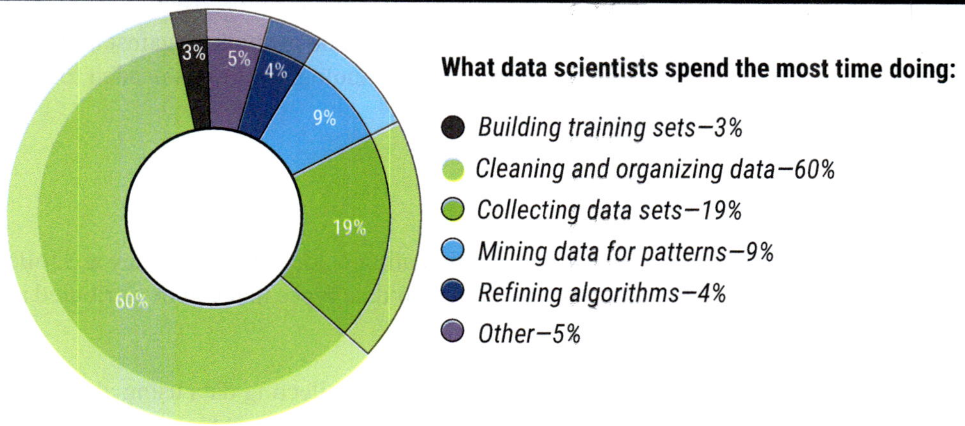

Source: *2016 Data Science Report*, Figure Eight (acquired by Appen in 2019), *https://visit.figure-eight.com/rs/416-ZBE-142/images/CrowdFlower_DataScienceReport_2016.pdf*

3.3.1 Database Queries

The database model dictates the type of algorithms that will be used for mining the data that reside in the data store. Relational data mining—also known as multi-relational data mining (MRDM)—is a data mining technique specifically for relational databases. It looks for the relationships, or contextual metadata of the relational tables (e.g., relationship between _Orders and _Customers). Traditional data mining algorithms use propositional patterns mining, which look for patterns in a single table, hence missing data dimensionality. The two differ slightly with respect to areas of use, approaches and classification rules.

Relational data mining is used when the data come from a data warehouse or from internal repositories such as ERP and CRM systems, where the data is highly structured and may require more of a translation—a mapping of sorts to remove duplication, level hierarchies and so on—to aid in proper classification.

Propositional patterns mining is traditionally used in machine learning and exploratory mining of semi-structured and unstructured data. Classification rules may be created as the data is analyzed, though careful uncoupling in structured data should be considered to avoid losing any of the relational data value.

NoSQL databases store data similarly to RDBMSes using a tables-and-rows structure. The formats and names of the columns vary from row to row across the table (also known as wide-column). NoSQL databases group columns of related data together. Related data can be retrieved in a single query operation, because only the columns associated with the search are retrieved. In a traditional relational database (RDBMS) the data is stored in different rows located in different places on a disk, requiring different queries and multiple disk operations for retrieval.

Moreover, although there is an ISO SQL Standard, ISO/IEC 9075:2016, it is not followed by all data owners, which further increases the preprocessing needed to prepare the data for analysis.

SQL implementations are incompatible between vendors—specifically with respect to date and time syntax, string concatenation, NULLs, and comparison case sensitivity—which certainly affects query design. Additionally, PostgreSQL does not adhere to the standard of folding unquoted names to uppercase, e.g., when a table is created with Real_Estate as the name. PostgreSQL would create an all lowercase table: real_estate. The tendency to default to all lowercase can make queries and all other data manipulations tricky, because developers would have to be very familiar with the tables and consistent in their queries.

3.3.2 Data Collection Methods and Techniques

Data acquisition can be performed in multiple ways. Each method can be evaluated against a few dimensions, such as cost, time to acquire, complexity, quality, skills required, and technologies or tools needed to evaluate the best approach to ensure highest-quality data.

Quantitative Data Collection

Quantitative data collection methods are straightforward, calculating cold, hard facts. They are collected mechanically and electronically. As such, they usually include no data types that can be attributed and expressed as broad human feelings and emotions.

Web scraping, crawling and streaming are programmatic methods of collecting data using algorithms. They can collect a quantity of data that far surpasses the volume of data that can be collected using qualitative data methods. A sample study that scraped the data of a Twitter user's account to infer personality from analyzing the user's tweets showed that the volume of data can be exponentially increased using the scraping technique (**figure 3.5**).

- **Web crawling** (uses crawlers or bots) is what most search engines do; they collect *all* data (structured and unstructured) without any intended data extraction goals defined. They use the information for search engine optimization and up-to-date indexing.
- **Web scraping** is targeted web analysis looking for specific data in order to extract it, usually structured.

Figure 3.5—Web Scraping Outperforms Crawling

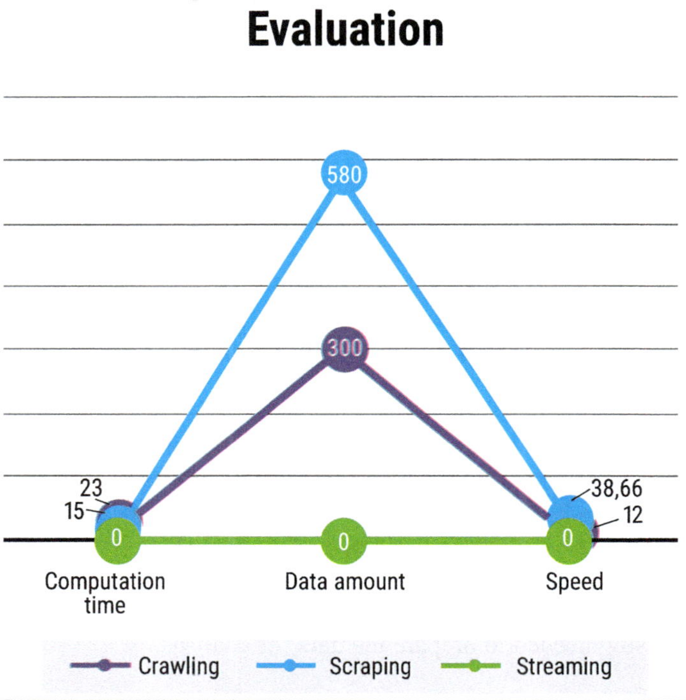

Source: Adi, S.; D. Hartanto; E. Utami; S. Raharjo; I. Oyong; "The Effect of Data Acquisition Techniques in Profiling Analysis Based on Twitter," *2019 International Conference on Information and Communications Technology (ICOIACT)*, 24 July 2019, DOI: 10.1109/ICOIACT46704.2019.8938549

Qualitative Data Collection

In contrast, qualitative data collection is more abstract and generalized. It may complicate and blur decision making and boundaries because it includes subjective assessments and collections. Qualitative collection methods can provide rich data about real people, including behavioral and emotional contexts. One weakness of these methods is that due to their nature, they may be difficult to reproduce.

Qualitative methods are costlier and more time-consuming. A task force of personnel is needed, and each individual is biased to some degree due to subjective thoughts, feelings and tendencies, whether consciously or unconsciously. Following are a few qualitative data collection methods:

- **Surveys**—Unlike many qualitative data collection methods, surveys can offer anonymity. It has been shown that people are more authentic when submitting opinions anonymously. Additionally, a survey allows for canned answers and easier categorization. The audience can simply select from the available options, producing results that can be summed up and reported.
- **Interviews**—Questions may be preselected and may be open- or close-ended questions. Interviewees may be led to select certain answers based on how interviewers word the questions. Open-ended questions can add value when the interviewee adds information the interviewer did not think to request.
- **Focus groups**—Focus groups are similar to one-on-one interviews with the difference of reduced time capital. One weakness of this method could be encouraging group think—a psychological tendency of individuals to choose the way the rest of the group does even when they think differently, to avoid standing out. This tendency can skew population data collection incorrectly.
- **Observation**—Observing participants' reactions, behaviors and nonverbal cues can capture information that a mechanical electronic machine never could. However, cues can be misinterpreted by observers, especially because cultural and global differences create variations in behavior and nonverbal cues.
- **Document/text analysis**—Information gained from reviewing scholarly reports, opinion articles, books, newsletters and publications depends on the reader's comprehension and ability to assess the information consumed. This type of data collection is extremely slow.

3.4 Data Classification

The data classification function includes all activities required to prepare data before it can be analyzed. The preparation step can be as short or as long as needed, depending on the data input sources and whether the data has been processed via an initial programmatic ETL process.

3.4.1 Data Cleansing

Before data cleansing can begin, the analyst must become familiar with the data to be cleansed and define what the structure, format, type, correct values, etc., should be, and the database that will use the cleansed data. Data cleansing may include any or all of the following:

Duplicated Records vs. Irrelevant Observations

1. Eliminate duplicate data records that may arise when the data is combined from multiple sources, clients, departments, etc.
2. Remove data records that are irrelevant to the project—e.g., if the analysis has to do with in-store purchases, all online sales should be removed.

Structural Errors

Structural errors will usually be encountered when the cleansed data is loaded to the application for analysis and modeling, so it is best to address them proactively.

1. Update all data to the correct data types (numerical, string, etc.)—e.g., "shingles" may be incorrectly classified as numerical.
2. All trailing and preceding spaces should be trimmed if not needed, especially if there are character limits.

Classification Errors

1. Update multiple similar values—e.g., health bar attribute values may show BAR, BAAR and BARS if the field is free form. All records that mean the same should be updated to conform to one consistent naming convention.
2. Sometimes acronyms are used and in other instances the words are spelled out. A consistent naming convention or a closed pick list may be necessary—e.g., N/A is the same as NA but would incorrectly create two classes instead of one.

NULL Values

1. Add missing values as placeholders if NULL data is not allowed into the system in which the data will be loaded (sometimes called imputing).
2. In some cases, NULL value *is the value*—e.g., "Automatic Transmission" field that is null indicates a no.
3. Thoroughly review records to determine whether to drop the records with mostly NULL data to avoid skewing results.

Other

1. Unify or strip formatting that is not needed.
2. Check for spelling errors but exercise caution—e.g., DISNY, DISNE and DISNEY may *not* all be the same, depending on context. Some apparent misspellings could be brand names and so should not be unified.
3. Check for grammar and correct capitalization when needed—this is especially important in branding.

It is a good practice to automate as much of the data cleaning as possible once the expectations and rules have been defined clearly.

The forthcoming sections will discuss varying algorithmic estimators that can be used depending on the type of data to be analyzed. **Figure 3.6** shows a high-level view of all of the algorithms and aids for choosing the appropriate algorithm.

Figure 3.6—Algorithms

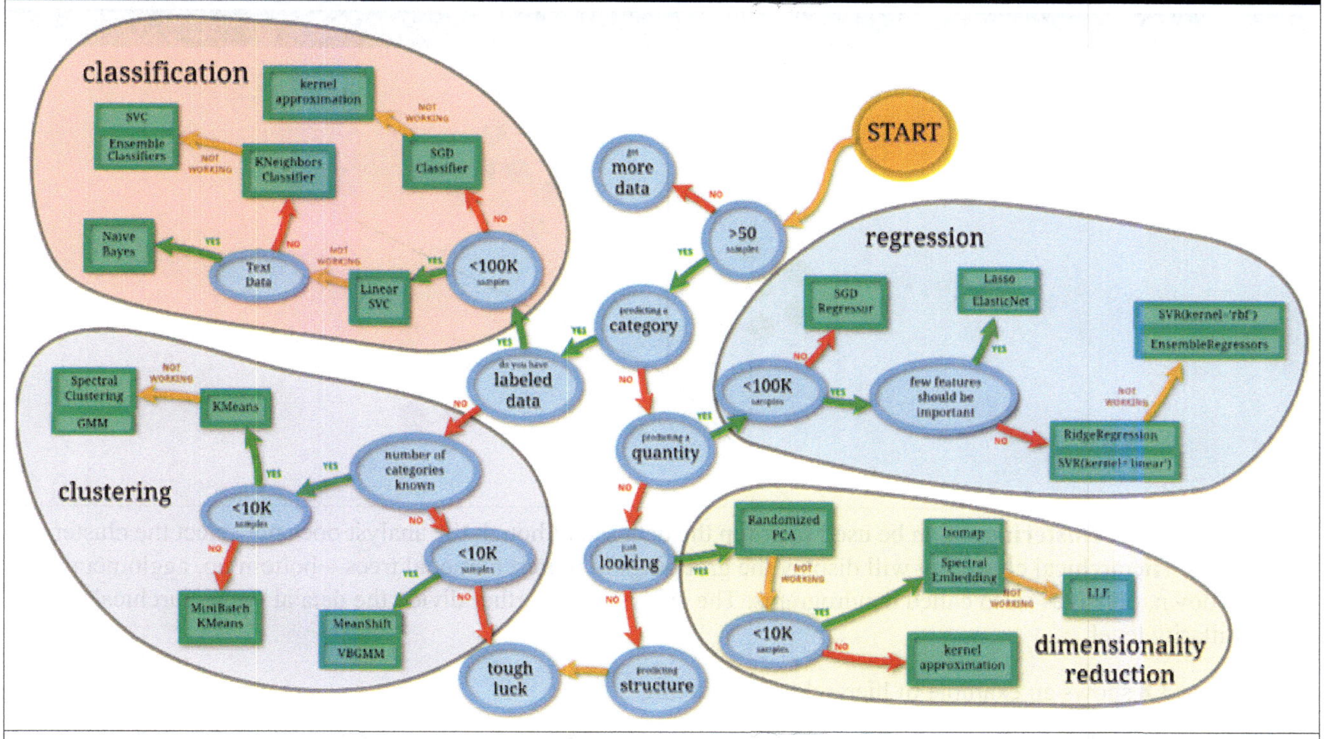

Source: Scikit-Learn, "Choosing the Right Estimator," *https://scikit-learn.org/stable/tutorial/machine_learning_map/index.html*

3.4.2 Data Clustering

Clustering is an algorithm for dividing data instances into groups—not a predetermined set of groups, which would make this classification, but groups identified by the execution of the algorithm because of similarities that it found among the instances. Clustering is a type of modeling that is unsupervised classification. It is not defined by the analyst in advance, possibly due to an unknown number of classes or class labels—e.g., the analyst may only know of roof materials that include asphalt and shingles and not all the variations.

Clustering is more advanced than ordinary classification where the data is well known. It is beneficial when used for pattern recognition, spatial data analysis, geographic information system (GIS) land analysis, insurance claims analysis and image processing.

Clustering methods are considered good if they produce intra-classes with data points of high similarity. They are considered lower quality when they produce inter-classes with data points of low similarity.

There are five different types of clustering methods:

- **Partitioning methods**—Partitioning methods divide the data into "K" clusters. "K" specifies the number of clusters specified by the analyst. "K" may be 5, 10, 4 or any number of clusters. There are different partitioning methods (K-means, K-medoids, CLARANS) but K-means is the most common. The center of each cluster, known as the "centroid," is the mean of the data points. This method is sensitive to outliers.

Figure 3.7 shows an example of a K cluster.

Figure 3.7—K-Cluster

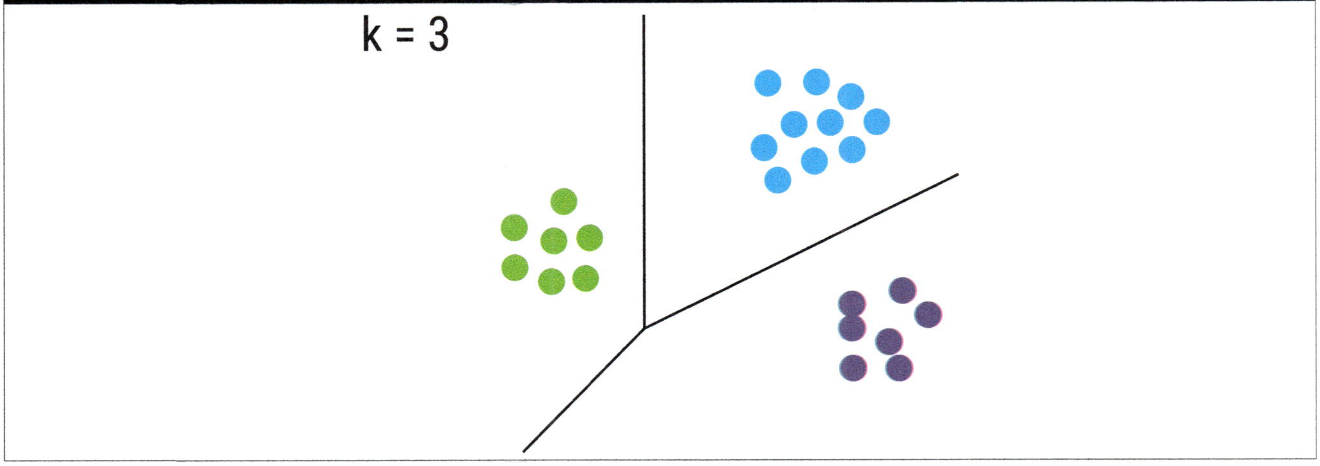

- **Hierarchical clustering**—Can be used to group the data set, although the analyst does not select the cluster counts. Hierarchical clustering will display the groupings in representational trees—bottom up, agglomerative; top down, divisive—also called dendrograms. The analyst may further divide the data at the hierarchical similarity level.

Figure 3.8 shows an example of hierarchical clustering.

Figure 3.8—Hierarchical Clustering

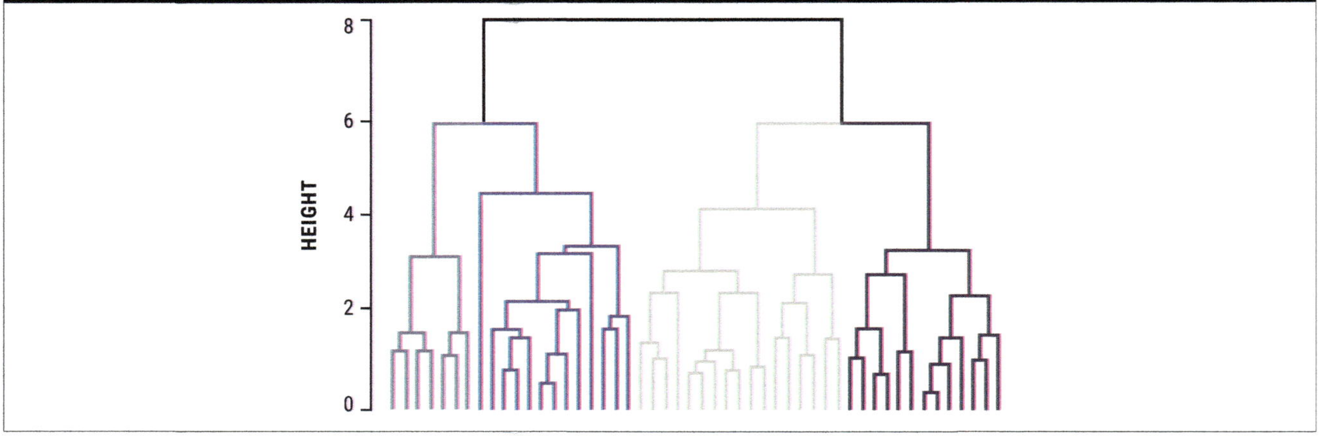

Other clustering methods are: BIRCH, CURE and CHAMELEON, although hierarchical methods do not scale well with voluminous amounts of data.

The following are considered advanced clustering methods:

- **Fuzzy clustering**—In fuzzy clustering, the same data can belong to more than one cluster and is dependent upon the coefficient that affects the degree of membership to each cluster (also known as "soft clustering"; K clustering is "hard clustering").

Figure 3.9 shows an example soft fuzzy clustering.

Figure 3.9—Soft Fuzzy Clustering

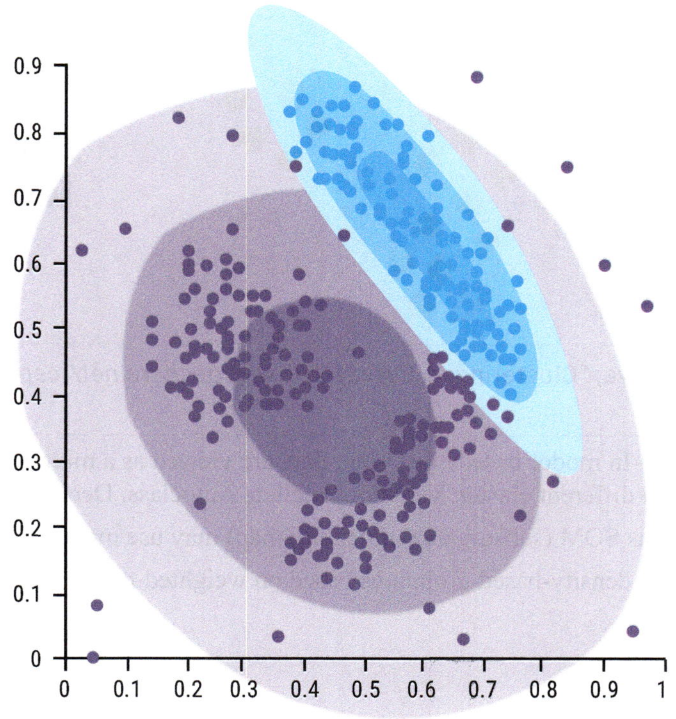

Source: Bozkir, S.; "FUAT - A Fuzzy Clustering Analysis Tool," *www.slideshare.net/aselmanb/fuat-a-fuzzy-clustering-analysis-tool*

- **Density-based clustering**—Density-based clustering is placing a number of data points within a radius calculation, EPS, which stands for optimal value for epsilon. It is especially useful for data with outliers and data noise. Variations are DBSCAN, CLIQUE and OPTICS.

Figure 3.10 shows an example of density-based clustering.

Figure 3.10—Density-Based Clustering

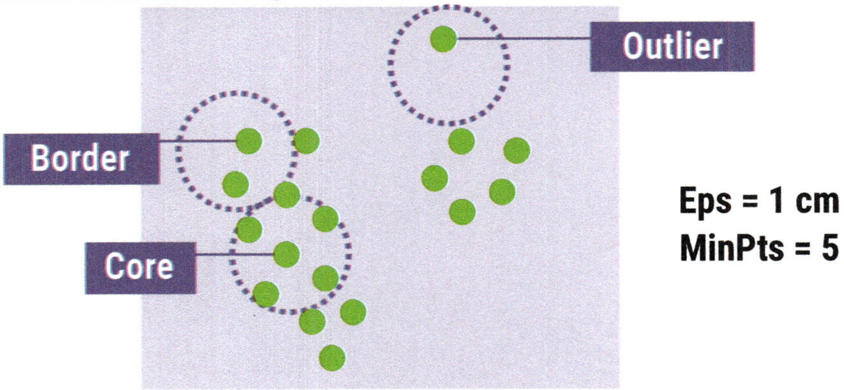

Source: Capaldo, R.; F. Collova; "Clustering: A Survey," *www.slideshare.net/rcapaldo/cluster-analysis-presentation*

- **Model-based clustering**—In model-based clustering, data are viewed as a mixture of probability distributions, each of which represents a different cluster. Variations include Autoclass, Denclue and Cobweb.
  - Neural network such as SOM (self-organizing feature map) may use model-based clustering.
  - Machine learning uses density-based groupings based on weighted feature sets.

3.4.3 Data Tagging

Most users are familiar with tagging social media posts using the hashtag symbol (#) to ease filtering and retrieval of information. However, this method created innumerable variants of the tag and there are no constraints around using a tag on a post or image of no relevance to the tag used.

In the data science context, organization of the data can be done via tagging, labeling or metadata. Taxonomy is a form of order arrangement and division of data in a framed structure that makes it easier to find when it is needed. Taxonomy organizes the data into a meaningful hierarchy, which is different from metadata and the usual tagging. Users often rely on taxonomies when shopping online and choosing different options for their selected product if a taxonomy has been created. Otherwise, they must resort to filtering products, which can make it difficult to find the right product in the shortest search time.

A few taxonomy best practices:

- Should have a logical hierarchy
- Should be easy to understand by nontechnical users in different business units
- Should not exceed five levels of nesting, if possible
- Should be media independent
- Should not be redundant to defined metadata
- Should avoid acronyms or abbreviations

Figure 3.11 shows an example of the Books section of an online retailer's taxonomy structure and a few examples of metadata that are visible. Tags and some metadata may be hidden, depending on the field, retailer and rules for each field (e.g., if field is null, it is not shown to the user).

Figure 3.11—Sample Taxonomy Structure

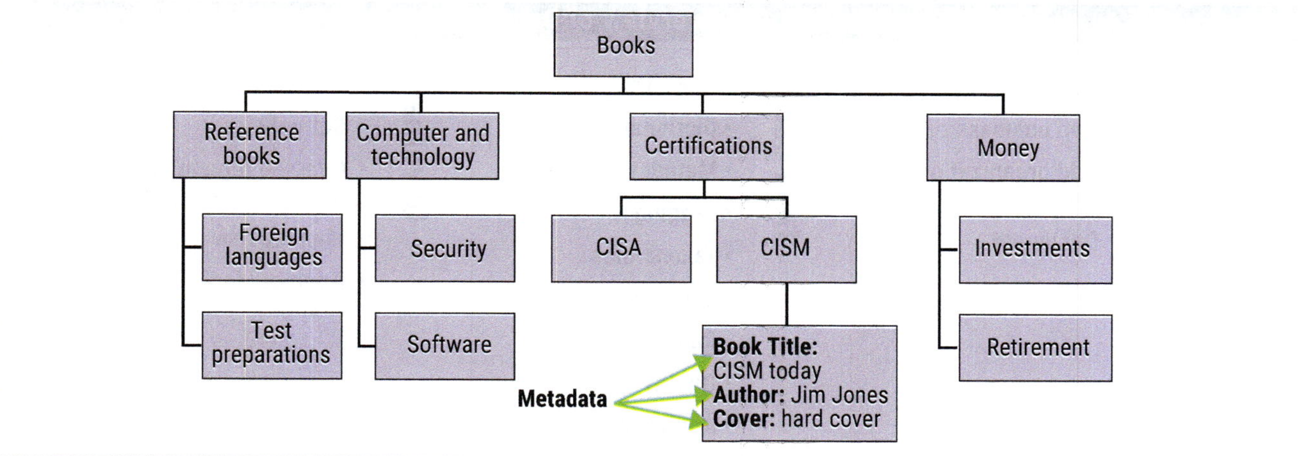

3.4.4 Data Governance Tools

Across the enterprise, there is a lack of visibility into what datasets exist. It is often difficult to understand how to access a dataset, and there may be no clear way to identify how a dataset has been used. Data governance teams use tools and an enterprise-wide data catalog to enable users to intelligently find, create, modify and manage the metadata and taxonomies of the datasets.

A data catalog is the detailed inventory of the data assets of an enterprise. The data connections, flows and business uses are curated and documented to assist analysts in understanding the data's purpose. A data catalog tool does not contain the data but rather the *metadata* about the data.

Some tools that are often used by mature organizations tasked with managing big data are: IBM, Collibra, Alation, Informatica and Azure Data Catalog, just to name a few. These tools allow data stewards to curate the data's accuracy—they are already responsible for collecting, collating, and evaluating issues and problems with data. The data steward's curation process may include ranking the data's accuracy from all sources, notating the business context, i.e., use cases, thus certainly increasing the data asset's value to the organization once it has been curated. After the curators supply data changes, the governance team may choose to approve or decline those changes. If approved, a new business rule may be implemented to automatically filter incoming source data, thus ensuring the data catalog (i.e., dictionary) becomes the decision gate for quality, trustworthy data.

Figure 3.12 shows an overview of a data catalog's use along with related tasks by governance and data stewards.

Figure 3.12—Overview of Data Catalog

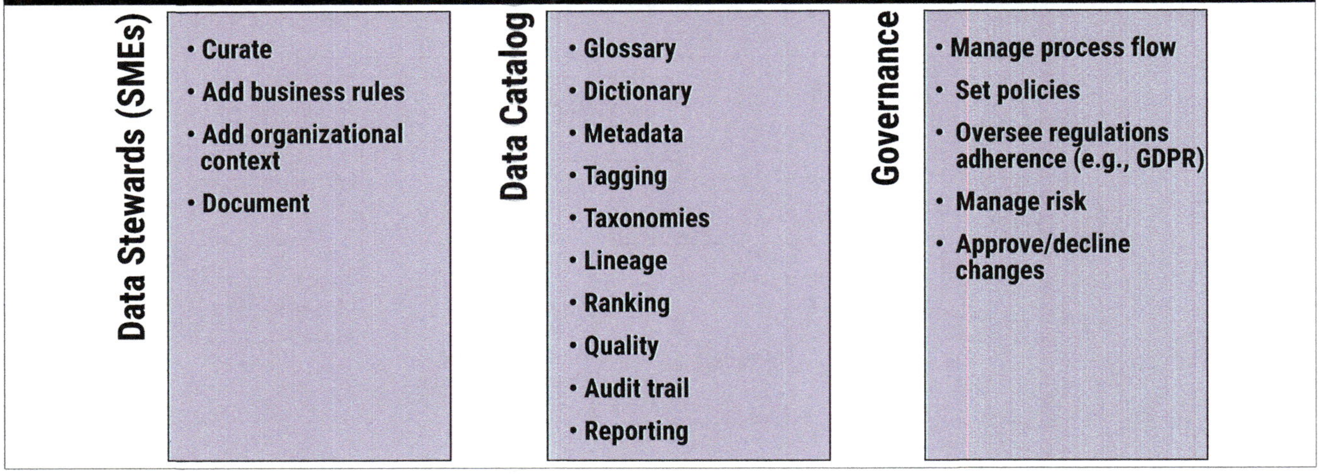

3.5 Data Processing

3.5.1 Introduction

Data processing includes several major tasks that collectively prepare the data for modeling post data collection. Processing usually includes understanding the data, massaging the data, uncovering relationships, and identifying dependent and independent variables, because data assets are multidimensional. This hands-on manipulation and visualization processing further refines the questions that were initially hypothesized.

3.5.2 Exploratory Data Analysis

Exploratory data analysis is a mostly graphical approach used to investigate the data in a curious manner; exploring the data structures, detecting underlying insights, and testing varying hypotheses. In statistical analysis, specific techniques are used to prove or disprove assumptions; however, in an exploratory approach, the data unveils the connections and context in the absence of preconceived assumptions. Analysts open-mindedly and inquisitively peruse data to intuitively and experientially discover patterns, spot anomalies, check assumptions and frame hypotheses.

Exploratory data analysis can be categorized into four high-level categories:

- **Univariate visualization**—Provides summary aggregates for each field in the raw data set.
- **Bivariate visualization**—Is performed to find the relationship between two (i.e., bi) variables, e.g., does a tree's age affect its growth pattern? The variables in the dataset are the tree's age and the target variable of interest, i.e., growth pattern.
- **Multivariate visualization**—Is performed to understand interactions between variables (typically three or more) in the dataset, e.g., comparing the popularity of four blog posts on a website, and then measuring their click rates for both Europe and North America to discover statistical relationships, if any.
- **Dimensionality reduction**—Is performed to simplify analysis. Two of the 5 Vs of data science are variability (i.e., variety) and sheer volume of data. These two factors can immensely complicate analysis due to the noise of variables and the inclusion of records that may not need to be considered. Exploratory data analysis helps the analyst to understand the fields in the data that account for the most variance between observations, and allow for the processing of a reduced volume and variance of data. It allows the analyst to become intimately acquainted with the data.

A few of the methods used are raw data plotting techniques such as:

- **Histograms**—Histograms can be used to answer questions such as:
  - What kind of population distributions do the data come from?
  - Where are the data located?
  - How spread out are the data?
  - Are the data symmetric or skewed?
  - Are there any outliers?[41]

  Tools such as spreadsheets and a general-purpose office suite can create most histograms, and they are often used in software development to describe bug counts, sprint velocity in agile, etc.

  Figure 3.13 depicts a histogram example.

Figure 3.13—Sample Histogram

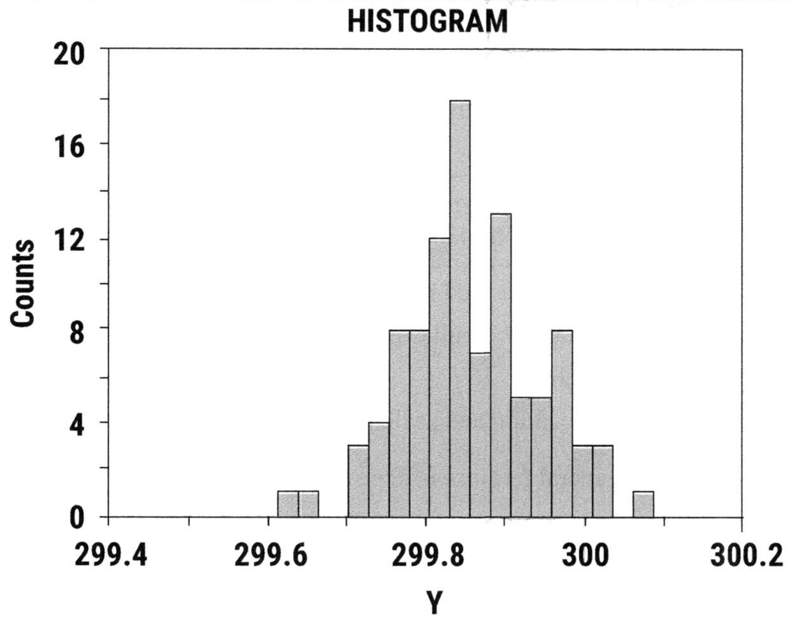

Source: *NIST/SEMATECH e-Handbook of Statistical Methods, www.itl.nist.gov/div898/handbook/eda/section3/histogra.htm*

- **Bihistograms**—Bihistograms can be used to answer questions such as:
  - Is a two-level factor significant?
  - Does a two-level factor have an effect?
  - Does the location change between the two subgroups?
  - Does the variation change between the two subgroups?
  - Does the distributional shape change between subgroups?
  - Are there any outliers?[42]

  Tools for creating bihistograms are not common, but they can be created in Dataplot software and using R.

[41] NIST Sematech, "Engineering Statistics Handbook," 1.3.3.14 Histogram, *www.itl.nist.gov/div898/handbook/eda/section3/histogra.htm*
[42] NIST Sematech, "Engineering Statistics Handbook," 1.3.3.2 Bihistogram, *www.itl.nist.gov/div898/handbook/eda/section3/bihistog.htm*

Figure 3.14 depicts a bihistograms example.

Figure 3.14—Sample Bihistogram

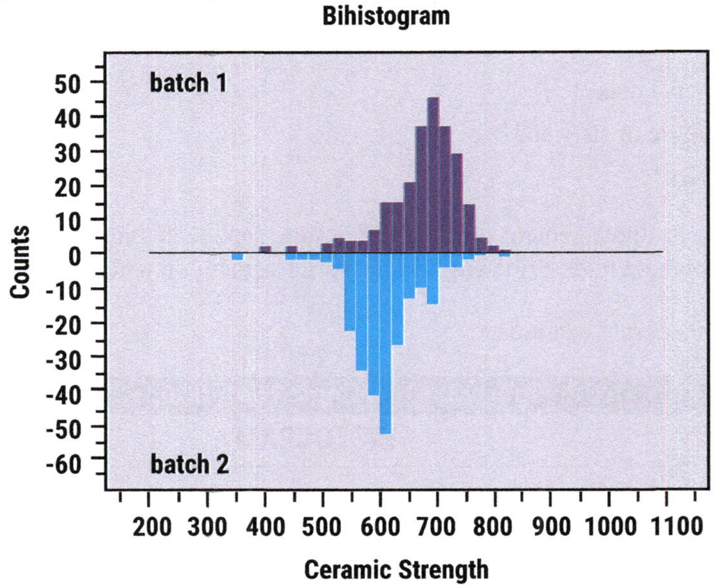

Source: *NIST/SEMATECH e-Handbook of Statistical Methods, www.itl.nist.gov/div898/handbook/eda/section3/bihistog.htm*

- **Probability plots**—Probability plots can be used to answer questions such as:
  - Does a given distribution, such as the Weibull,[43] provide a good fit for the data?
  - What distribution best fits the data?
  - What are good estimates for the location and scale parameters of the chosen distribution?[44]

Tools such as spreadsheets and a general purpose office suite can create most probability distribution plots.

[43] For more information on Weibull, see Statistics How To, "Weibull Distribution and Weibull Analysis," *www.statisticshowto.com/weibull-distribution/*

[44] NIST Sematech, "Engineering Statistics Handbook," 1.3.3.22 Probability Plot, *www.itl.nist.gov/div898/handbook/eda/section3/probplot.htm*

Figure 3.15 depicts a probability plot example.

Figure 3.15—Sample Probability Plot

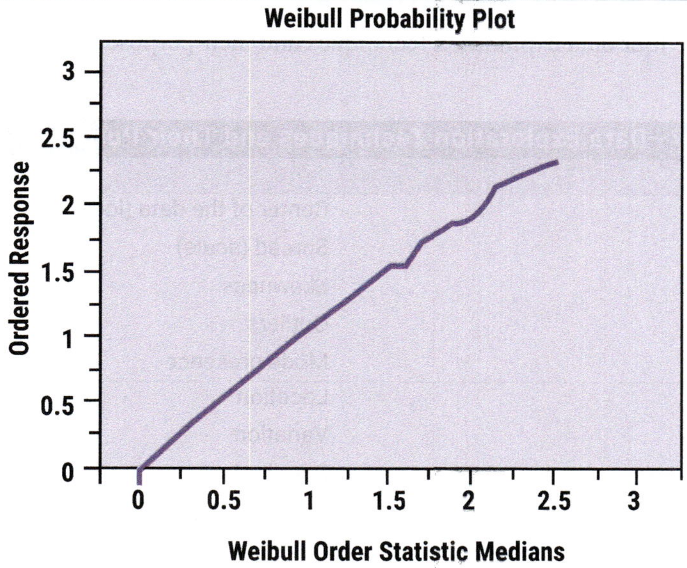

Source: *NIST/SEMATECH e-Handbook of Statistical Methods, www.itl.nist.gov/div898/handbook/eda/section3/probplot.htm*

- **Data trace**—Also known as run-sequence plot, a data trace is similar to lag and scatter plots and can help answer the following questions:
  - Are there any shifts in location?
  - Are there any shifts in variation?
  - Are there any outliers?[45]

Tools such as spreadsheets and statistical software (e.g., SAS) can create most data trace and scatter plots.

Figure 3.16 depicts a run-sequence plot example.

Figure 3.16—Run-Sequence Plot

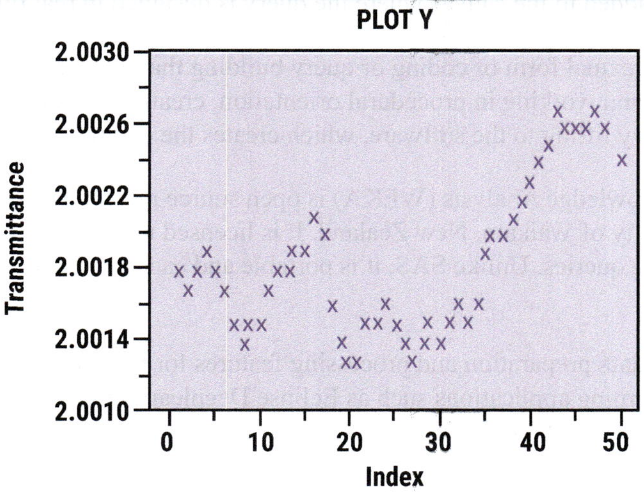

Source: *NIST/SEMATECH e-Handbook of Statistical Methods, www.itl.nist.gov/div898/handbook/eda/section3/runseqpl.htm*

[45] NIST Sematech, "Engineering Statistics Handbook," 1.3.3.25 Run-Sequence Plots, *www.itl.nist.gov/div898/handbook/eda/section3/runseqpl.htm*

These techniques can be used concurrently to expand on each other's findings and check for correctness. For example, using a probability plot can test the distributional model of a histogram.

Figure 3.17 summarizes the four data exploration techniques and their purposes.

Figure 3.17—Data Exploration Techniques and Their Purposes: A Summary

Exploratory Analysis Technique	Graphically Denotes
Histogram (Univariate)	Center of the data (location) Spread (scale) Skewness Outliers Mode presence
Bihistogram (Bivariate)	Location Variation Distribution
Probability Plot (Multivariate and dimensionality reduction)	Distribution of data flow: plotting correlation coefficient
Data Trace Run-Sequence Scatter Plot	Random drawings Fixed distribution Common locations Common scale

Source: Adapted from *NIST/SEMATECH e-Handbook of Statistical Methods, www.itl.nist.gov/div898/handbook/eda/section3/runseqpl.htm*

3.5.3 Model Development Tools

The SAS® Enterprise Guide® Query Builder is one of the most powerful components of the commercial SAS software. Although SAS uses SQL it has a graphical user interface with drag-and-drop functionality to connect input data sources, join tables and columns, sort, filter, change data types, and much more, without having to write a single line of SQL.[46] The application integrates the SQL code on the back end (hidden from analyst accessible) as the entities and connections are added to the canvas where the query is designed in real time.

Rather than using a verbose, textual form of coding or query building that requires experience, the application caters to those who prefer thinking and working in procedural orientation, creating a visual representation of the data structure and leaving the heavy lifting to the software, which creates the coding.

Waikato Environment for Knowledge Analysis (WEKA) is open source machine learning software written in Java and developed at the University of Waikato, New Zealand. It is licensed under GNU General Public License and rivals SAS's GUI for building queries. Unlike SAS, it is portable and can be used on the three major platforms: Windows, Linux and macOS.

Weka offers a wide range of data preparation and processing features for both relational structured data and flat files. It integrates with machine learning applications such as Eclipse Deeplearning4j programming library, which incorporates Java Virtual Machine (JVM) portability to extend to any platform.

Deep learning is a branch of machine learning that mimics neurons and neural networks in the brain and is associated with thinking in human beings. Deep learning can be used for speech recognition, translation and image recognition.

[46] For further description of the SAS® Enterprise Guide® Query Builder, see First-Kluge, J.; S. First; "The Query Builder: The Swiss Army Knife of SAS Enterprise Guide," *https://support.sas.com/resources/papers/proceedings14/1557-2014.pdf.*

Weka has both a GUI and a command line interface and from its workbench, known as a canvas or integrated development environment (IDE) in other applications, it can perform the following:

1. Tree modeling
2. Regression
3. Neural network modeling for multilayer visualization
4. Visualization for scatter plots and clusters

MATLAB (Matrix Laboratory), created by MathWorks, is a language and an interactive environment for developing algorithms, analyzing and visualizing data, and performing numerical computations (**figure 3.18**). It supports data input such as flat files, other applications or external devices (such as Raspberry Pi).

MATLAB is used to teach linear algebra in academia, for numerical computing in engineering and for image processing in the scientific world. Its programming language is proprietary. Although the software requires a license (available at a reduced cost for students) it can add a wrapper to use C and the older FORTRAN language. It offers a robust graphing capability using mathematical equations and supports development of a GUI.

Figure 3.18—MATLAB

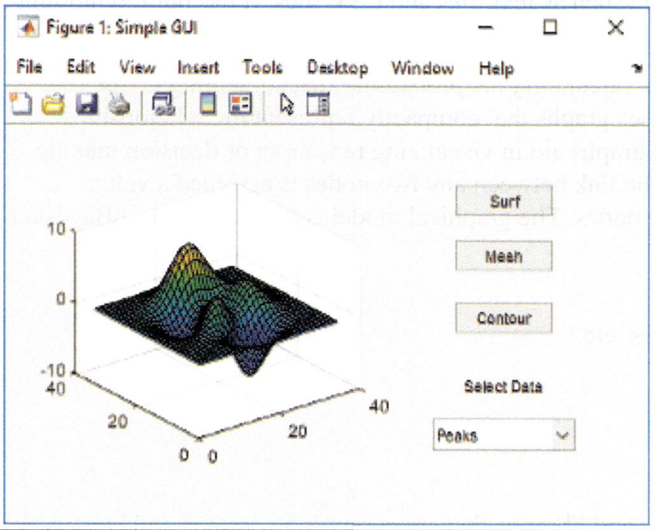

Source: MathWorks, "Create and Run a Simple Figure-Based App," *www.mathworks.com/help/matlab/creating_guis/about-the-simple-programmatic-gui-example.html.* MATLAB R2020b, The MathWorks, Inc., Natick, Massachusetts, USA.

Alpine Miner is a data modeling application that enables business analysts with no development or programming experience to create analytic workflows and predictive models. It offers a graphical user interface with a drag-and-drop design. It integrates well with Oracle and other RDBMSes, R programming language, and Hadoop, and it can access data from any device. Unlike other tools, Alpine Miner includes enterprise-level collaboration and governance features, reinforcing the idea that data science is an enterprisewide collaborative effort.

Python is an object-oriented programming language that is free to use for commercial or personal projects, and it is relatively easy to learn for those without a programming background. It is often used in data science, because its community development resulted in an extensive collection of tools. One of its benefits is the inclusion of Pandas, an open source library for data manipulation and analysis that rivals Excel's feature set and is capable of handling very large datasets.

3.5.4 Statistical Analysis Tools

Statistical tools visualize mass amounts of data and assist organizations in identifying patterns faster than a team of individuals could ever do manually. Some of the following software applications are mature in feature sets and algorithms, further expediting any exploratory analysis.

SAS

SAS is a statistical software suite developed by the SAS Institute for data management. It is used for advanced analytics, multivariate analysis, business intelligence, criminal investigation and predictive analytics in commercial enterprises, the financial industry, academia and government.

It can be used both on Windows and Linux platforms. It is written in the C programming language and has a robust GUI. It utilizes procedural SQL and Python's scripting language to write algorithmic procedures used for querying and manipulating input datasets from a relational database with structured data. SAS can be used with structured RDBMSes including Excel, MS Access, IBM's DB2, Teradata and Informix.

It offers powerful querying capabilities on unstructured data such as text files and CSV files. It has built in plotting charts and canned reports that accommodate the following techniques:

1. **Bayesian analysis**—Bayes' theorem is an equation for calculating the probability that something is true if something potentially related to it is true. It encompasses graphs that compactly represent the relationships between random variables for a given problem. These graphs aid in visualizing reasoning or decision making. Networks are usually represented as graphs in which the link between any two nodes is assigned a value representing the probabilistic relationship between the nodes. The graphical model is also known as a Bayesian network.
2. **Descriptive and predictive analysis**
3. **Slew of mixed models** (categorical, inferences, clusters, etc.)
4. **Missing data imputation**

R

R is a programming language and a tool that uses packages and libraries that are community-created and loaded. For those with a fair level of programming experience, it is useful for writing lengthy code for new procedures, or modifications to existing shared procedures.

R is an open source solution that offers most of the features and capabilities of SAS without the high cost of licenses needed when using SAS. SAS, owned by a large enterprise, offers support, expert training, documentation and, most importantly, robust security features. In contrast, R relies on community forums. R saves licensing cost, but it may require a longer learning curve. R is constantly being updated. Created in 1993, it is not as advanced as SAS, which was created in the 1970s.

Some of R's benefits:

1. Programming constructs and flexibility for conditional, user-defined functions, loops, etc.
2. Ecosystem of users and other open source tools
3. Advanced deep learning support

Figure 3.19 shows Google trend interest in R and SAS over the last 16 years. As R continues to gain market share and improve its security and feature sets, it may become a serious contender to SAS and an essential tool in the data science domain.

Figure 3.19—Google Trend of Searches on SAS and R

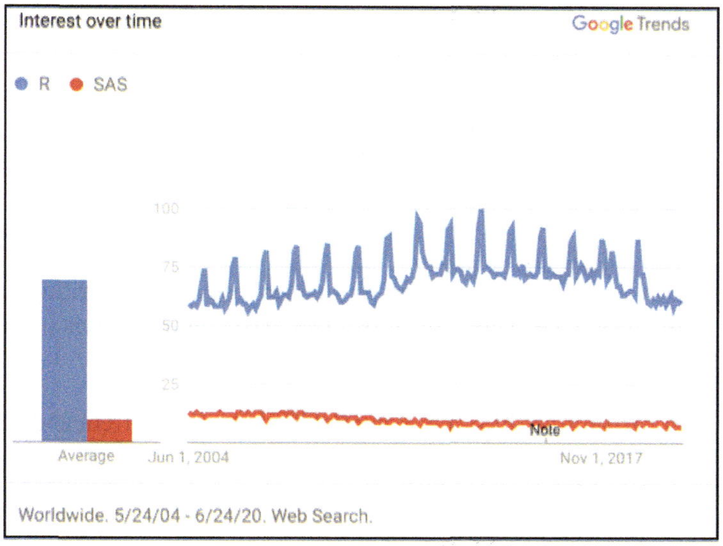

Source: Google Trends, *https://www.google.com/trends*

3.5.5 Business Analytics

Organizations choose to implement business analytics and use data science to obtain the capabilities to analyze historical data and forecast the future in order to make better decisions that result in improvements such as optimized sales channels, informed training programs, focused marketing campaigns, lowered operating costs. They also use these tools to improve customer service and modify product offerings.

To make good decisions, decision makers need input in response to the following questions:

1. What happened?
2. What could happen?
3. What should we do?

There are three distinct analytical techniques that help ask and answer those questions.

Descriptive

Descriptive analytics asks historical questions: What happened? To what degree? Why? What was the result of what happened? Did we meet our goals? If not, what were the consequences?

The data to be analyzed is first aggregated from sales and financial reports, NPS, etc. It is cleansed, if needed, and mined for insight and patterns. The business metrics are reviewed, as descriptive analytics relies heavily on KPI governance and whether goals were initially set to understand what the descriptive data analysis reveals when the two are compared. Without KPIs and a governing body, analytics will likely play a marginal role in decision making.

Once prepared, the data is analyzed via aggregation and summary reports, and compared to KPI metrics and other goals to measure performance. Finally, the data is presented via graphs and charts to decision makers.

In summary, descriptive analysis is a surface view of the visible, available data from the past. It does not attempt to infer or predict; it is able to compare past years' reports and identify a downward (or upward) trend.

Predictive

Predictive analytics asks forward-looking questions in an effort to understand the future. It uses predictive algorithms to provide actionable insights based on data that an organization can implement in the present. It can employ "what if" scenarios, and it examines past historical data to estimate the likelihood of certain outcomes based on probabilities.

One current example of predictive analysis is insurance companies' use of underwriting procedures to qualify customers for life, auto or home insurance. They use consumers' past claim histories (or average life spans and habits, for life insurance) to assess the risk of having a claim filed.

Another common example is a person's credit score. Potential lenders view current and past payment histories and credit scores to determine the probability that applicants will make future payments on time. Credit bureaus provide risk scores that aid lenders in making their determinations.

Prescriptive

Prescriptive analytics come into play when stakeholders who have received descriptive and predictive analysis results ask for recommendations—i.e., what can we implement now to get the results outlined in the predictive phase?

In the prescriptive phase, optimization guidance is added to the scenarios simulated with the predictive technique—e.g., if we combine product A (our best seller) with product B (our median seller) in gift packs during November and December, we can increase the price of Product A and move more inventory of Product B while still increasing the profit margin by 1.2 percent compared with last year. Next year's Product B inventory should be reduced by 34 percent.

These suggestions are actionable and can be visualized, quantified and virtually prototyped before implementation actually occurs. Stakeholders can discuss, fine-tune and change product options to view new options and probabilities.

Prescriptive analytics applies business rules, statistics and machine learning algorithms to internal and, if desired, external data—such as social media profiles—to make impactful suggestions.

In some texts, a fourth technique may be named independently, although it is usually performed as part of the descriptive technique and, if separated, it will follow the descriptive technique. As the descriptive technique examines what happened, the diagnostic technique asks *why* it happened and looks for the underlying root cause.

No one technique is better than the others. The recommendation is to use all three, as they complement each other, in order to create a holistic view of the enterprise's past market performance and how it can compete efficiently within that market by implementing prescriptive recommendations.

3.5.6 Machine Learning

Machine learning is a program or system that builds (trains) a predictive model from input data. The system uses the learned model to make useful inferences from new (never-before-seen) data drawn from the same distribution as the one used to train the model. It adjusts its behaviors based on feedback from data.

Machines use algorithms adaptable to the circumstances once they have been trained to predict outcomes without explicit human input. As more data is processed by the machine to gain experience, it is able to use statistical analysis to adjust and update behavior to more accurately predict the future.

There are three main machine learning types:

- **Supervised learning**—Develops predictive models based on input and output of available data
  - Algorithms used
    - Classification (e.g., Naïve Bayes, nearest neighbor, support vector machines and discriminate analysis)
    - Regression (e.g. linear regression, decision trees, neural networks)
- **Unsupervised learning**—Discovers internal representations from input data, no predictions
  - Algorithms used
    - Clustering (e.g., K-means, hierarchical, neural networks, etc.)
- **Reinforcement learning**—Learns from input data and makes decisions from output data to confirm lessons
  - Algorithms used
    - Decision making (e.g., Q-learning, R-learning, TD-learning)

The learning mechanisms and algorithms most often used for predictive models are classification and regression. Classification algorithms teach the machine how to predict which class the data will be classified into based on specific characteristics the machine algorithm can look for and identify. For example, a classification algorithm can be shown images of different snakes initially to capture characteristics such as length, coloring, etc., and then the algorithm can be presented with a new image of a snake to make a classification decision on its own.

At the point when the algorithm makes its decision, the reinforcement learning idea is applied. Should the algorithm classify the image properly, according to its training, a reward is provided (e.g., a point) as a positive reinforcement. If the algorithm makes an incorrect classification (e.g., snake image is classified as tree branch), a punishment ensues for negative reinforcement (e.g., the algorithm loses the game or a point is subtracted).

Applying a confusion matrix can help reduce, or at minimum identify, some algorithmic classification errors. A confusion matrix can test classification algorithms for accuracy. Accuracy can be reduced when there are more than two classes (e.g., snake, cat and dog) or unequal number of algorithmic observations (e.g., seven images).

Figure 3.20 shows a confusion matrix using the animal classification example.

Figure 3.20—Confusion matrix

Test Runs: N = 5575	**Predicted:** Not a Snake	**Predicted:** Snake	**Totals**
Actual: Not a Snake	**True Negatives** = 1052	**False Positives** = 1399	2451
Actual: Snake	**False Negatives** = 497	**True Positives** = 2627	3124
Totals	1549	4026	5575

The model was used to run through 5,575 (N=5575) test runs to see if the algorithm could classify the image cards correctly as either Snake or Not a Snake (actually Tree Branch).

- **True negatives (TN)**—This number signifies the actual number of times the algorithm predicted the image as ***not*** being a Snake correctly.
- **True positives (TP)**—This number signifies the actual number of times the algorithm predicted the image as a Snake correctly.
- **False negative (FN)**—This number signifies the actual number of times the algorithm predicted the Snake image as being a Tree Branch incorrectly.

- **False positive (FP)**—This number signifies the actual number of times the algorithm predicted the Tree Branch image as a Snake incorrectly.

Applying the following equation will reveal the algorithm's classification accuracy:

(TP+TN)/Total Predictions * 100 (2627+1052)/5575*100 = 65.99% accuracy

At 66 percent accuracy, this algorithm can use some more training and likely some programming updates. To calculate the other rates:

- **Error rate**—(FP+FN)/Total Predictions * 100

  (1399+497)/5575 * 100 = 34%
- **True positive rate**—(when it is a Snake it actually predicts Snake): 2627/3124= 0.84
- **False positive rate**—(when it is a Tree Branch it actually predicts a Snake): 1399/2451=0.57
- **True negative rate**—(when it is Not a Snake it actually predicts it is Not a Snake): 1052/2451=0.43
- **False negative rate**—(when it is a Snake it actually predicts it is Not a Snake): 497/3124=0.16

Regression algorithms predict what number a variable will have, e.g., predict next year's sales. If the variable changes over time (e.g., gas price increases) the regression is called a time-series prediction.

In some cases, as the algorithm learns it may continuously make bad decisions and become what is considered a weak learner. This produces a weak prediction model in the form of an ensemble of weak prediction models, typically decision trees. The gradient-boosting correction technique builds the model in a stage-wise fashion, like other boosting methods do, and it generalizes them by allowing optimization of an arbitrary differentiable loss function.

Regardless of the specific algorithm selected, the results are dependent on how the training models are designed and whether lessons learned during the model test runs were incorporated in the training model iterations. Here are a few principles to consider:

1. **Sample size of training data**
2. **Training data types (preprocessed or not processed)**—In some cases, preprocessing or classifying the data may actually create a constraint for the model and result in less accuracy and precision, requiring preprocessing model adjustments.
3. **Missing data generalizations**—Consider how the algorithm should decide when there is missing null or partial data (explained in section 3.4).
4. **Feature (attribute) selection**—Analyze combinations of attributes in the data to find which subset of features works best for prediction. This technique removes data noise (i.e., complexity) and allows the machine to train faster and improve accuracy.

Naïve Bayes

Naïve Bayes, from the Bayes Theorem, is the simplest of all the Bayesian probability techniques—hence the naïve reference. Naïve Bayes is a collection of algorithms that share a common principle: that every feature being classified is independent of the value of any other feature (attribute weight is equal). Hence, its accuracy is not the most reliable, except in document and text classification where it excels.

Naïve Bayes notation: **Pr**(C|d) where **Pr** = probability, (C|d) = "d" belongs to "C" class

K-Nearest Neighbor (K-NN)

K-NN is used when the data distribution knowledge is very little or unknown. The algorithm makes a best-case classification based on the nearest neighboring record.

The K-NN algorithm is effective with large datasets, simple to implement and not noise-sensitive. However, it is very sensitive to irrelevant or redundant features, because all features contribute to the similarity and thus to the classification. Two other disadvantages of the method:

- With distance-based learning it is not clear which type of distance to use and which attributes to use to produce the best results.
- Computation cost is quite high due to the need to compute the distance of each query instance for all training samples.[47]

When K-NN is used for regression, the predicted value is the average value of its neighbors. When using this algorithm for regression prediction it is best to use the weight dimension, assigning weight to distance. The further the neighbor, the less weight it has; the closer it is, the more it affects the final predicted value (see **figure 3.21**). The algorithmic measurement for the distance is usually one of the following metrics: Euclidean, Euclidean squared, City-block, and Chebychev.

Figure 3.21—K-Nearest Neighbor (K-NN) Example

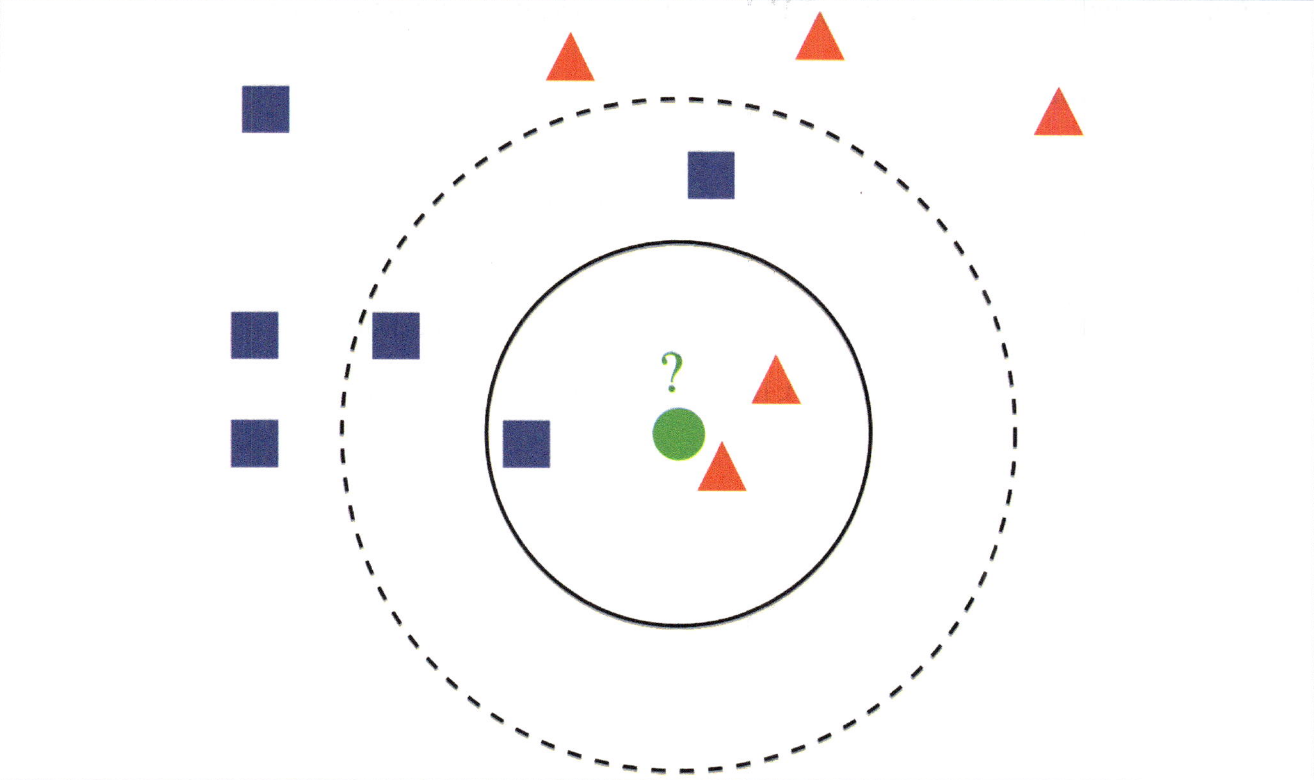

Source: Wikimedia Commons, "KnnClassification," *https://commons.wikimedia.org/wiki/File:KnnClassification.svg*. This file is licensed under the Creative Commons Attribution-ShareAlike 3.0 Unported license.

[47] Imandoust, S.B.; M. Bolandraftar; "Application of K-Nearest Neighbor (KNN) Approach for Predicting Economic Events: Theoretical Background," *International Journal of Engineering Research and Applications*, vol. 3, issue 5, September-October 2013

The new data point, the green circle shape, needs to be classified properly as either Category B or Category A by the algorithm. The K-NN algorithm can use several calculation metrics to determine the distance of the neighbors (**figure 3.22**).

Figure 3.22—K-Nearest Neighbor (K-NN) Euclidean Distance Calculation

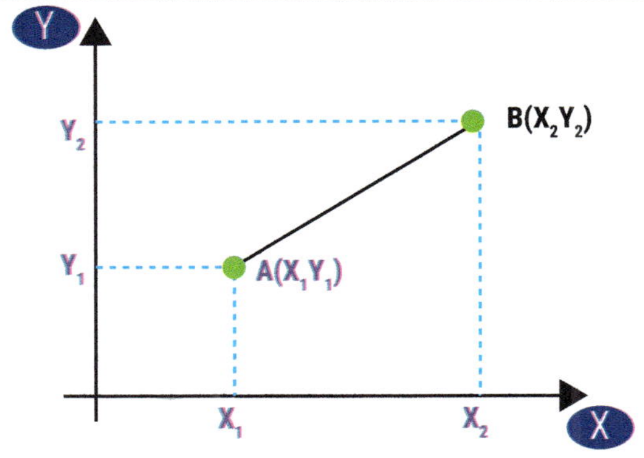

Euclidean Distance between A_1 and B_2 = $\sqrt{(X_2-X_1)^2+(Y_2-Y_1)^2}$

In this example, the Euclidean calculation signifies the distance between two points—the shorter the distance the more likely the classification. Moreover, notice in **figure 3.21** that Category B has two closest points, while Category A only has one. *Which category does the green circle belong to with this added information?*

Support Vector Machine (SVM)

SVM is an algorithm that uses hyperplanes to maximize the distance from one data classification point to the nearest data classification point. This allows the algorithm to weigh the distance between the classes based on the one that represents the largest separation.

Figure 3.23 illustrates SVM; there are three hyperplanes: H_1, H_2, and H_3. The distance between each class (black dots is one class; white dots is another) is measured in its proximity to the hyperplanes.

H3 does not separate the 2 classes (green line); H1 (purple line) and H2 (blue line) do. H_1 with a small margin and H_2 with the maximum margin.

Figure 3.23—Support Vector Machine Hyperplanes

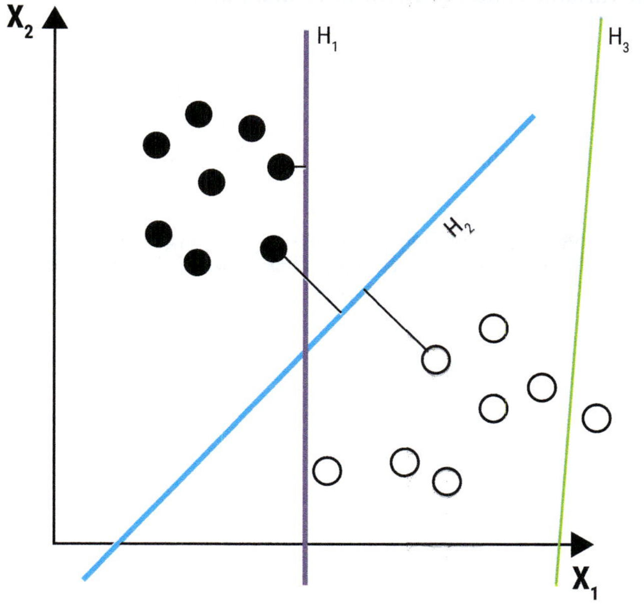

Source: Wikimedia Commons, "Separating Hyperplanes," *https://commons.wikimedia.org/wiki/File:Svm_separating_hyperplanes.png*

Feature Selection

Picking up on the home prices example in prior chapters, note that square footage is a dimension of a home's price, but other dimensions—such as city, state, demographics, employers in a 10-mile radius, proximity to shopping and many more—are added dimensions (also known as features, variables, attributes, etc.) that can affect the analysis (and create noise) if not carefully chosen. Dimension reduction is a term used to describe the transformation of high dimensionality data to lower dimensionality data while still retaining meaningful representational attributes of the original data. In other words, it is a technique to extract one or more dimensions that capture as many variations in the data as possible, also called projection.

For example, **figure 3.24** shows all possible dimensions reduced from eight to four dimensions.

Figure 3.24—Data Dimensionality

Dimensions	Dimensions Reduction
Square footage Number of bedrooms Number of bathrooms	Size
City State ZIP code	Location
Employers in a 10-mile radius Proximity to shopping	Employers in a 10-mile radius Proximity to shopping

An applicable technique used to reduce dimensionality is principal component analysis (PCA). PCA presents a linear representation of the principal variables and the covariance mapping.

Assuming each dimension is a column in a table, there are too many columns in this case (consider dimension reduction for 10 or more features) to process and analyze them at once. **Figure 3.25** shows the data mapped with two projection lines. A projection line is used in this case: If a line is drawn next to mapped data, which projection line (i.e., perceptions or dimensions) will capture the widest range of the data and map it sparsely across the line? See **figure 3.26**.

Figure 3.25—Data Dimensionality Projection 1

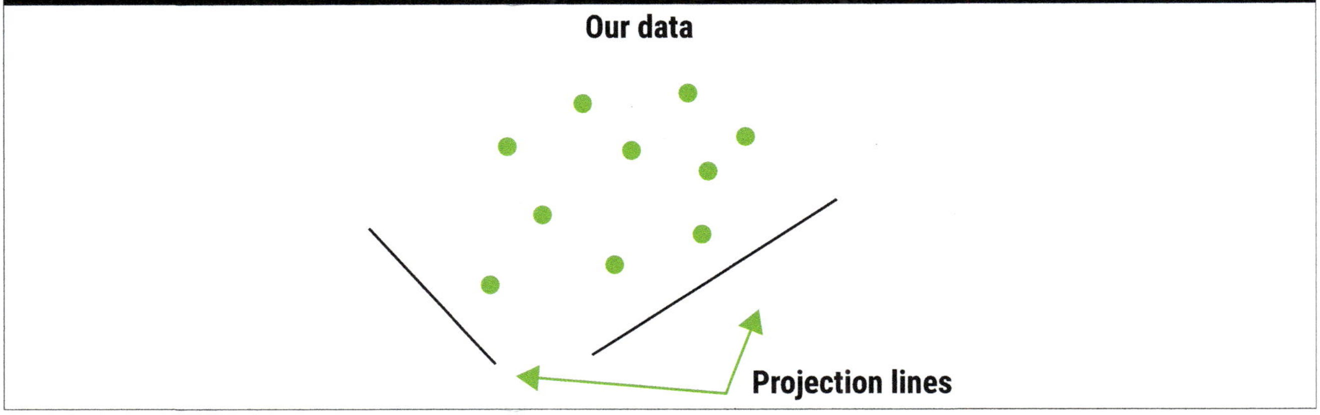

Figure 3.26—Data Dimensionality Projection 2

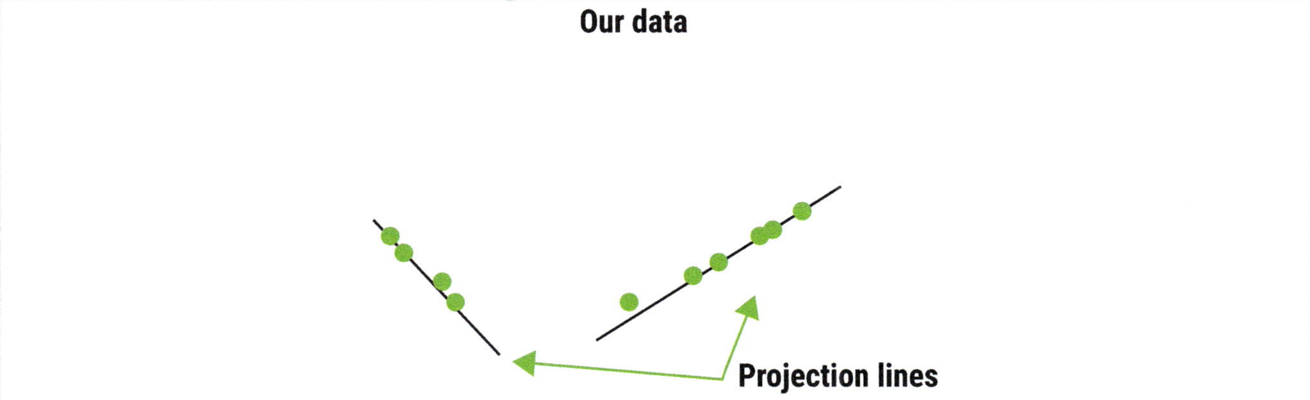

PCA calculations help the analyst decide which features to keep based on the linear spread, variance and covariance of the data points. PCA's eigenvectors notate the direction of the linear line (positive, negative or zero/low covariance, **figure 3.27**), and eigenvalues inform of the magnitude of the features.

Figure 3.27—Covariances

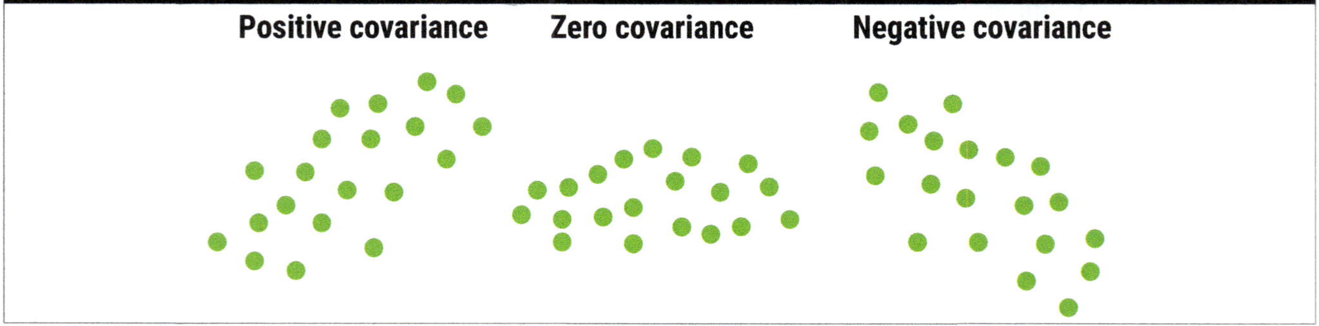

An eigenvector is a nonzero vector whose direction changes at most by a scalar factor when linear transformation is applied to it. The eigenvalue is the factor by which the vector is scaled. A positive covariance signifies the variables have a relationship of moving together, i.e., affecting each other. A negative covariance signifies an opposing relationship. Zero covariance signifies no, or very low, covariance.

Neural Networks

Consider the complexity of the human brain and the ability to identify patterns and learn from experiences. Habits, whether considered good or bad, are only as strong or as weak as an individual's neuron connectivity:

- Neuron switching time: ~.001 seconds
- Number of neurons: $\sim 10^{10}$
- Connections per neuron: $\sim 10^{4-5}$
- Scene recognition time: ~0.1 seconds
- Mostly parallel computations[48]

A machine algorithm made to mirror human neuron activity can be considered weak or strong based on its decision-making abilities, accuracy and efficiency. Neural networks are highly parallel processing activities that are multilayered and nonlinear, made possible by complex mathematical models and formulas.

Neural networks have three layers:

- **Input layer**—The input layer passes the data (i.e., input) in raw form without processing or computational tasks.
- **Hidden layer**—The hidden layer is a conduit between the input layer and the output layer, where all the computational work is performed.
- **Output layer**—The output layer receives the output of the processing done in the hidden layer.

Although the computational tasks are different the idea is the same as software development's IPO (Input, Process, Output) process (**figure 3.28**)

Figure 3.28—Overview of the IPO (Input, Process, Output Cycle

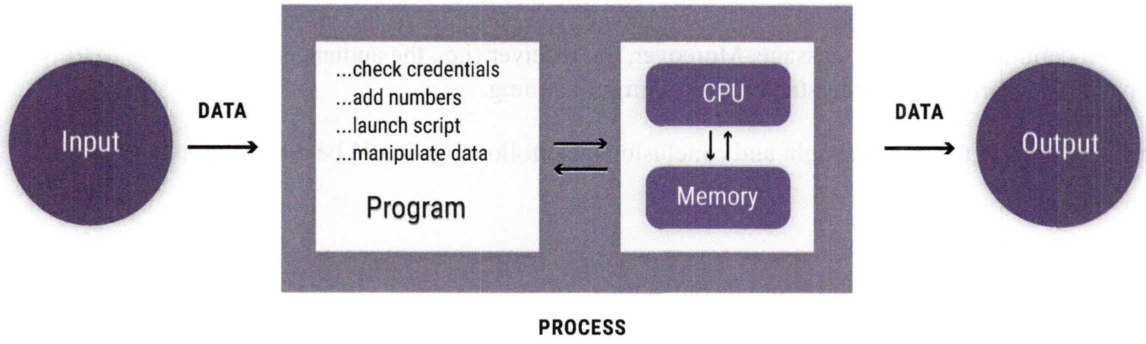

Source: ISACA, *Software Development Fundamentals Study Guide*, USA, 2021, figure 2.1, *https://www.isaca.org/bookstore/bookstore-misc-ebook/witassd*

[48] Mitchell, T.M.; "Chapter 4: Artificial Neural Networks," *Machine Learning*, McGraw-Hill Science/Engineering/Math, USA, 1997, *https://dl.matlabyar.com/mousa/Machine%20Learning%20matlabyar/e-book/McGrawHill%20-%20Machine%20Learning%20-Tom%20Mitchell%20(better).pdf*

Neural networks can be shallow (one level) or deep (greater than two levels). A neural network is comprised of several elements:

- **Perceptron**—Responsible for the algorithms for supervised learning of binary classifiers. A binary classifier is a function that can decide whether or not an input, represented by a vector of numbers, belongs to some specific class.[49]
- **Neural net**—Consists of thousands or even millions of simple processing nodes—also known as the connectors between the perceptrons—that are densely interconnected and used for carrying out computations.
- **Sink node**—Refers to the deepest level of a node, from the input layer all the way to the output layer, as shown in **figure 3.29**.

Figure 3.29—Figure Reference

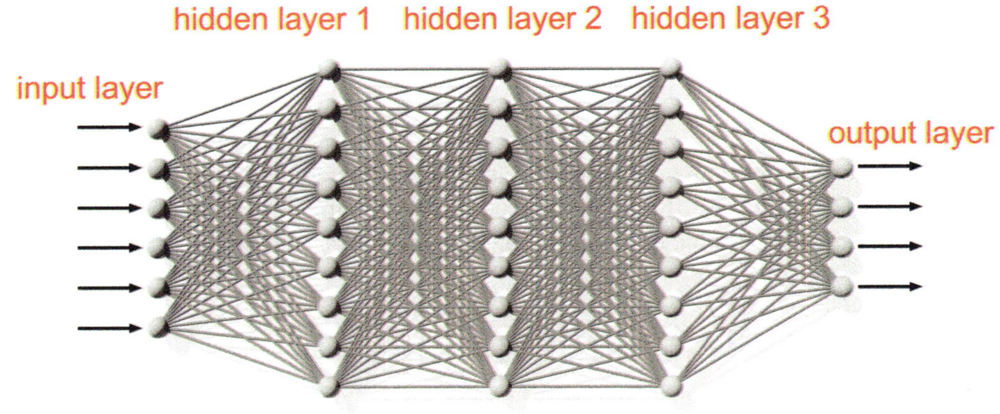

Source: Eugenesergeev, iStock, Getty Images

3.6 Communication of Results

One of the three pillars of a data scientist's skillset is the ability to communicate data results to enterprise stakeholders. Data scientists, engineers and analysts excel at the technical side of big data, but unless they are able to communicate findings in lay terms and business lexicon, the data's message can be lost in translation. Communication, by definition, means that the sender (i.e., the presenter of the information) has to be clear, concise and direct in communicating the message. Moreover, the receiver (i.e., the audience) must be able to truly comprehend the information. Understanding is a form of learning.

In any plan to present the data's insight and conclusions, the following should be considered:

1. Know the audience
2. Make no assumptions
3. Speak in business language and avoid technicalities
4. Compare and contrast results to business goals and governance KPIs
5. When prescribing actions, be specific and expand on why
6. Allocate time for Q & A

[49] Freund, Y.; R.E. Shapire; "Large Margin Classification Using the Perceptron Algorithm," *Machine Learning*, 37(3):277-296, 1999, *http://cseweb.ucsd.edu/~yfreund/papers/LargeMarginsUsingPerceptron.pdf*

3.6.1 Reporting Techniques

The old adage "a picture is worth 1000 words" still holds true today. Considering that different people learn and understand in different ways, it is worthwhile to incorporate the following learning styles in communicating results:

- **Auditory**—Presenting the information and speaking to it using the audience's lexicon will help tremendously in articulating findings.
  - Every data point, regardless of subject or industry, should be tied back to an organizational goal, and hence to one of its department's internal, dependent goals—change the message associated with the same data to fit the audience.
  - Create a narrative of the story the data asset wants to share and the insights uncovered. For example: The data reveal that when a salesperson calls a prospective client an average of 3.5 times pre-sales, 95 percent of sales close within 45 days vs. 97 days when pre-sales calls are two or fewer.
- **Visual**—Visualizations using graphs, charts, numbers and summaries allow the audience to reach the same conclusions quickly and retain information longer. Although all individuals have a preference for the way they learn, all seem to appreciate visual images. One word of caution: Visualizations can be misleading and lead to incorrect interpretations.
  - Aesthetic design principles and elements—such as color, shape, labeling, size, ordering, comparison, etc.—can have a big influence on how people interpret the data.
- **Kinesthetic**—Although it may not be possible in all scenarios and business contexts, it may be worthwhile to include a few *what if* scenarios that can be used during the presentation to get the audience engaged, prompt questions, and build trust in the results, encouraging collective buy-in. Specific reporting tools (explained in section 3.6.2) allow the audience to interact with the data, ask their own questions, and apply filters that are especially important to their business need or role.

Figure 3.30 shows an overly complex, difficult-to-understand data visualization. In this chart, the graph shows too much data, and thus any insights to be gleaned are lost in this chart.

Figure 3.30—Data Visualization: Deficient Example

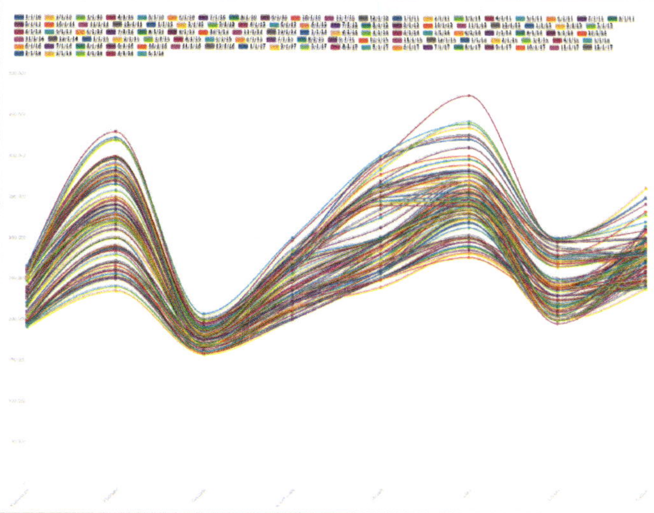

Source: Learning Solutions, "Misleading Data Visualizations Can Confuse, Deceive Learners," *https://learningsolutionsmag.com/articles/misleading-data-visualizations-can-confuse-deceive-learners*

Following is a listing of visualization charts that specialize in data and information visualization. Histogram, scatterplot, tables, pie chart and line chart are omitted, given that they are covered in section 3.5.2 or are well known.

The following data visualization types are used for displaying quantitative data in schematic form (with or without axes):

- **2D Cartesian Coordinates**—Visualization is best used to display four different points of data using quadrants and X, Y coordinates. An example of this visualization type is a well-known strategic organizational portfolio matrix known as the Boston Consulting Group (BCG) Matrix. The BCG is often used when an enterprise evaluates the performance of its products and services, and for optimizing the allocation of funds.

  In this example (**figure 3.31**), the market share and market growth of the enterprise's products are measured along the X and Y axes respectively. Each quadrant represents the following:

  - **Stars** signify best performing products and highest market share
  - **Question marks** signify a product that has low market share in a high-growth market (based on data collected). In this case, given the market is growing quickly, additional funds would need to be diverted to this product (e.g., Cheese) to exploit the market growth, and improve sales and, eventually, market share.
  - **Cash cows** are products that produce cash.
  - **Dogs** are products are in negative market growth, so there is no need to invest additional funds.

Figure 3.31—Data Visualization: BCG Using 2D Cartesian Coordinates Visualization

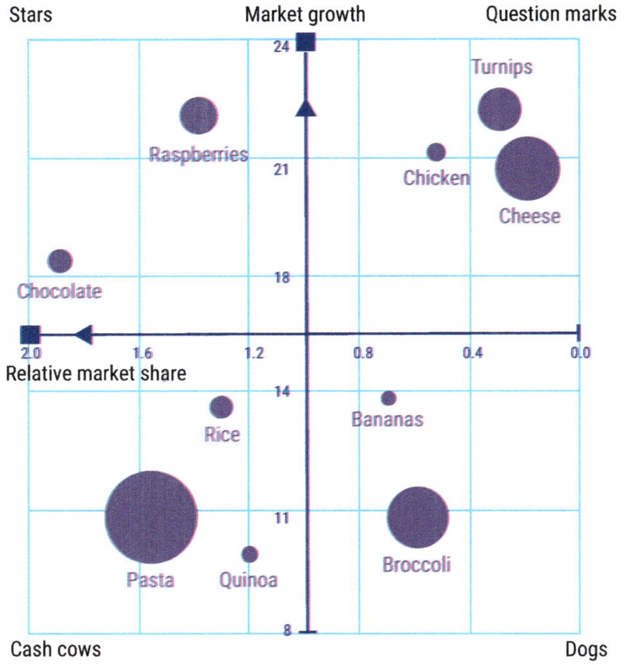

- **Spectrograms**—Visualization is used for audio, speech and voice pattern recognition to identify spoken words phonetically. It can show frequency and seismic signal strength of earthquakes and other events. Spectrograms are two-dimensional graphs with a third variable depicted by colors signifying another data dimension (**figure 3.32**).

Figure 3.32—Data Visualization: Seismic Data Activity Using Spectrogram Visualization

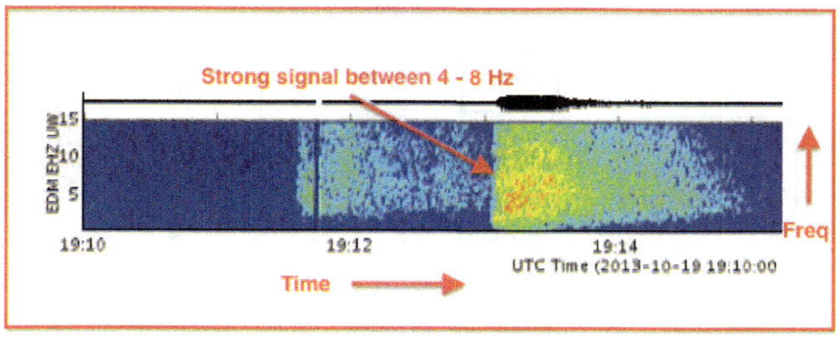

Source: Pacific Northwest Seismic Network, "What is a Spectrogram?," *https://pnsn.org/spectrograms/what-is-a-spectrogram*

The following example shows one visualization using a Sankey diagram (**figure 3.33**) and another using a bubble chart (**figure 3.34**). The information in both is exactly the same, and nothing was omitted. Chart selection alone made a difference in comprehension by the audience. The purpose was to relay the organizational chart of the House Democrats' health plan. Take note of the colors, shapes and connotations in **figure 3.33**. The human subconscious assesses patterns and makes connections almost immediately, creating an emotional response. *What do the shapes mean? The colors?*

Figure 3.33—Information Visualization: Organizational Chart of the House Democrats' Health Plan

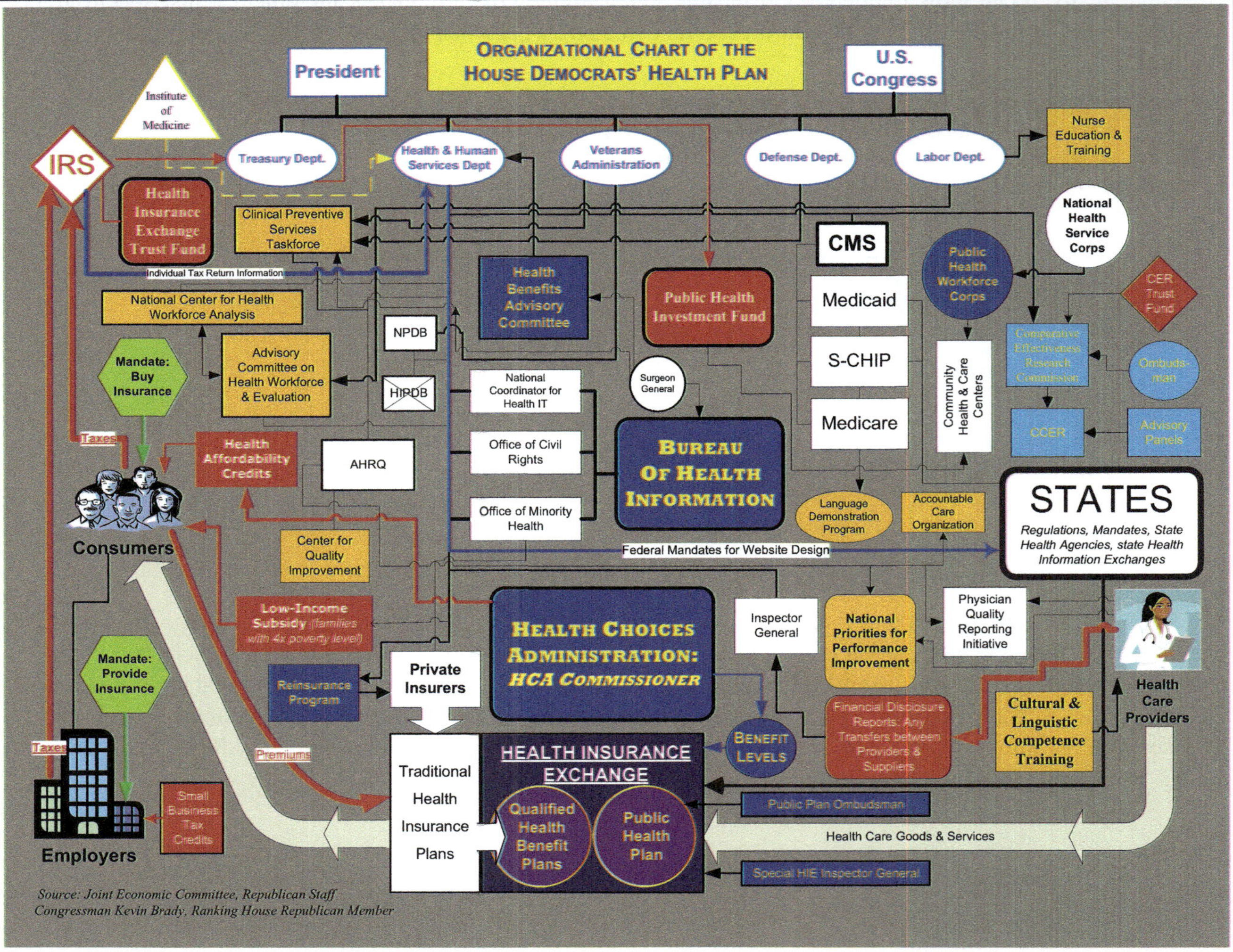

Source: Website of Rep. Don Young, *https://donyoung.house.gov/uploadedfiles/house-democrats-health-plan.pdf*

The same information was visualized by a digital designer named Robert Palmer who thought the information was presented entirely too complex and created the following replacement shown in **figure 3.34**. *What do the shapes mean to you? The colors?*

Figure 3.34—Information Visualization: Redesigned Organizational Chart of The House Democrats' Health Plan

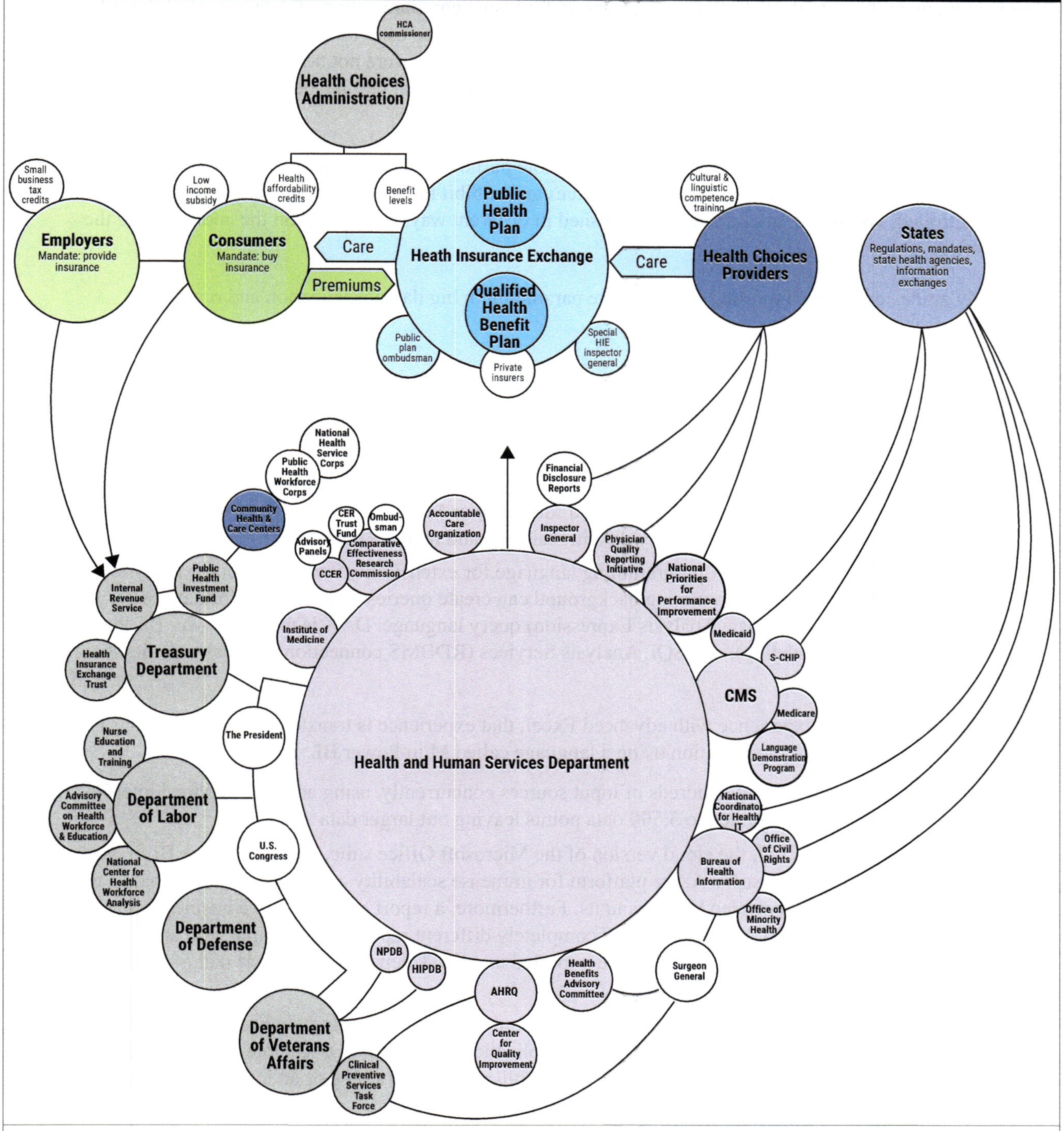

Source: Palmer, R.; Revised design of "Organizational Chart of the House Democrats' Health Plan," *https://www.flickr.com/photos/robertpalmer/3743826461/*

3.6.2 Reporting Tools

There are many analysis tools that offer canned reports for visual communication of data as part of the software. Additionally, generally used office products, such as Excel, can also be used for histograms, scatter plots, area charts and other simple, commonly used charts. The visualization feature sets were not necessarily the initial intent of the software, however, so they may not be as feature rich, abundant in digital graphic design options or dynamic (most are static) as needed to communicate results successfully.

In data science, there are several reporting tools whose sole purpose is to visualize data. They handle mass integrations of data flowing from multiple input sources and exhibit the data in a centralized dashboard that can be access controlled, visualized, interactive and published in different ways depending on the audience and the visualization.

This study guide covers two reporting tools that are paramount in big data visualization and reporting.

Microsoft's Power BI

Microsoft's Power BI (Business Intelligence) is part of Microsoft's ecosystem of applications offering interactive business analytics, predictive modeling, reporting and optimization (**figure 3.35**).

- **Ease of use**—Easy enough for non-programmers to create a quick dashboard using Excel (one of Power BI's greatest integrations, importing Excel's advanced formulas) or MS-Access data, incorporating charts or creating reports for publishing. It does not classify data in real-time and requires data classification pre-visualization.
- **Languages**—Can be used with R programming language for extensibility with open source analytics. Developers and those with programming background can create queries and data transformations for further customization using DAX (Data Analysis Expression) query language. DAX is used in Power BI, Power Pivot (drag-and-drop data modeling), and SQL Analysis Services (RDBMS connection and real-time dashboard updates).

  For users with lots of experience with advanced Excel, that experience is transferable. The functions are also transferable with a bit of modification using a language called M in Power BI.
- **Input data**—It supports many hundreds of input sources concurrently, using an easy graphical interface for connectors. However, it is limited to 3,500 data points leaving out larger data sets.
- **Cloud or local**—Office 365, the cloud version of the Microsoft Office suite, includes Power BI. It can be local or cloud based, using Microsoft's Azure platform for immense scalability and sharing of dashboards and reports across the enterprise and between business units. Furthermore, a report created by one business unit (BU) can be used as input for another BU's dashboard and completely different report. For example, a finance department salary and human resources cost report can be imported into an IT department's report that includes human resources cost as just one dimension.
- **Integrations**—Microsoft's longevity and history in the information technology industry has allowed it to create innumerable integrations with its own and others' applications, resulting in a scalable application that can be quite portable and shareable across industries (e.g., publish reports to a blog or an intranet SharePoint site). It connects to Azure-based stream analytics and Azure machine learning for real-time dashboards.
- **Platforms**—Dashboard (reports and visualizations) designers have to use two applications: Power BI Desktop, supported only on Windows platforms, is used to cleanse data, develop queries, design charts, create input connections, etc. The Power BI service application is accessible via a browser, as it is hosted in the cloud. The service application allows the designer to create or pin charts on other dashboards, control user access, schedule automatic data refreshes directly from input sources and configure notification schedules sending the reports to users' inboxes. Audience members can interact with the charts, ask questions of the data, and create their own reports via filtering and sharing.

- **Cost**—Microsoft's Power BI requires a license to publish dashboards to the web, whether internal or external. However, the desktop application can be used as part of an Office 365 subscription, and PDF or spreadsheet exports offer a way of sharing while not publishing.
- **Security**—Security is integrated with the access control needed for normal enterprise operations already in place when the office suite is used, and with Azure Active Directory (AAD) when the service is used for publishing. Designers can create roles and allow role-specific access for dashboards, dictating the features available to each role in addition to the enterprise security.

Figure 3.35—Data Reporting Tools: Microsoft Power BI

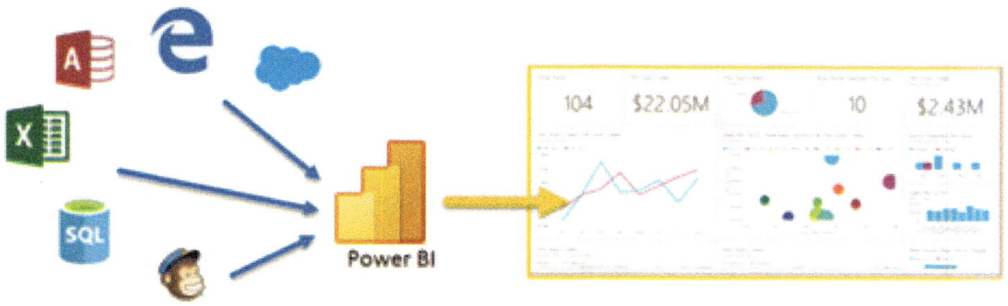

Source: Microsoft, "Power BI Weekly Service Update," *https://docs.microsoft.com/en-us/power-bi/consumer/end-user-reading-view*

Tableau Software

Tableau Software is a newer company than Microsoft. Founded in 2003 in California, it was later purchased by Salesforce, a CRM company. Tableau is a great alternative to Microsoft's BI and boasts that creating dashboards can be done with no coding or programming experience (**figure 3.36**). Aside from data visualizations, it offers metadata management and vertical and horizontal scalability.

- **Ease of use**—It requires no programming experience at all to create a quick dashboard and visualizations with many built-in templates to choose from. For developers, Tableau integrates well with R programming language, extending its capabilities to include predictive and statistical analytics.
- **Input data**—It supports Excel, PDF, Oracle or Amazon Web Services (AWS), and classifies data, hence reducing prep time and enabling the "ask questions" feature in the interface during dashboard creation.
- **Cloud or local**—Tableau Online, one of Tableau's applications, is its cloud version. It offers a full feature set, not partial, as is the case with Power BI's service cloud version.
- **Integrations**—It integrates with SQL Server, Mongo, Oracle, IBM DB2, Hadoop and others. With the additional TabPy Python scripting language, it can be used to add machine learning libraries to extend visualizations.
- **Platforms**—Tableau's ecosystem consists of:
  - **Tableau Desktop**—Installed locally on Windows only, it can be used to design the graphs and input data from any heterogeneous structured sources, such as data warehouses or Excel spreadsheets.
  - **Tableau Server**
  - **Tableau Reader**
  - **Tableau Public**—It enables free publishing of the created dashboards and reports.
- **Cost**—Tableau requires a license to use and publish and is costlier than Microsoft's product.

- **Security**—Tableau Server uses SSL/TLS to encrypt transmissions from clients to Tableau Server and from Tableau Server to the input sources, such as databases. Moreover, it offers much more granular security by implementing row level security (RLS), restricting specific rows of workbook data from defined users.
- **Scalability**—It can be scaled vertically, as it offers a multithreaded server/client architecture, and it can scale horizontally, as it is multithreaded and capable of integrated clustering technology.

Figure 3.36—Data Reporting Tools: Tableau Server's Architecture

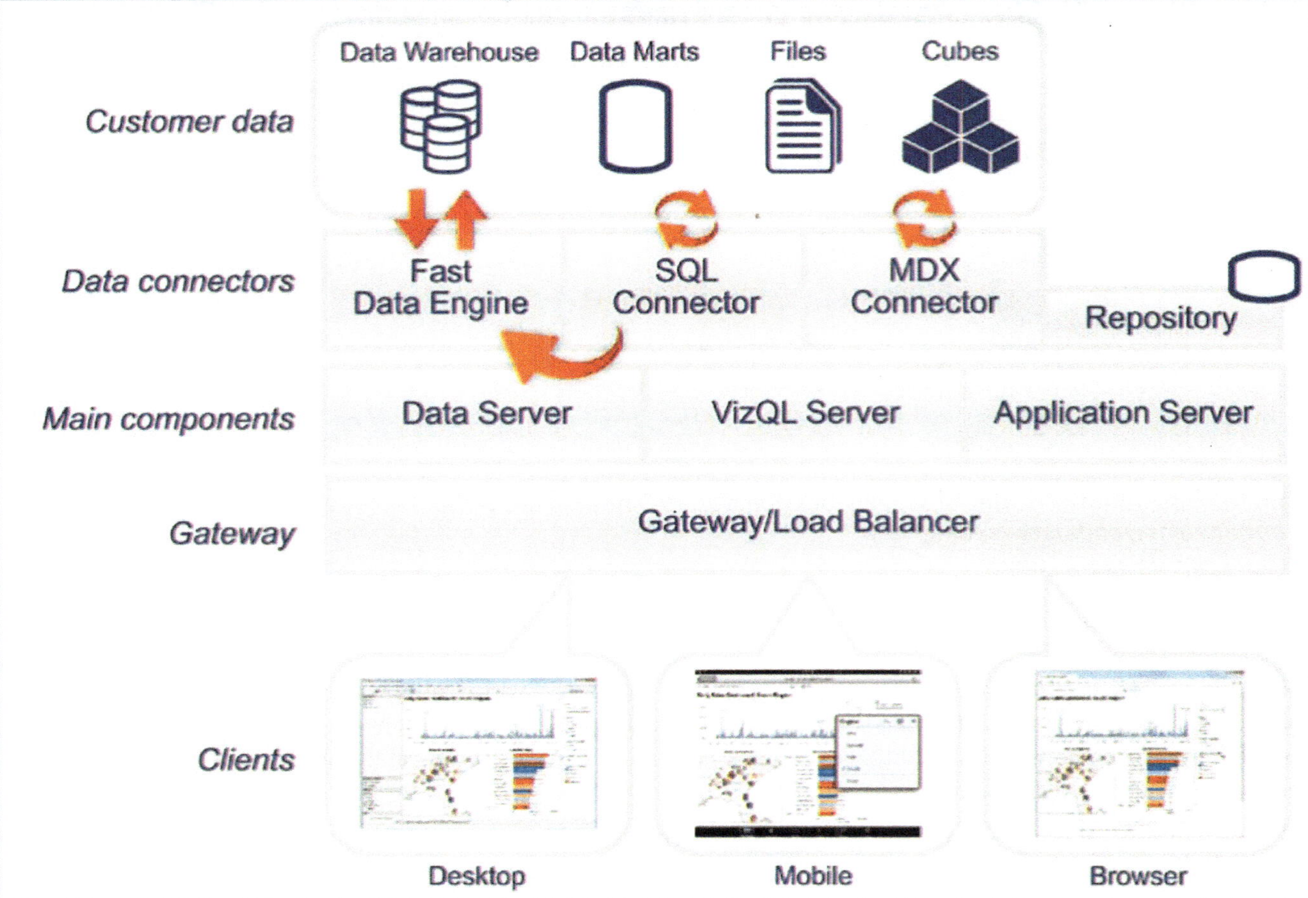

Source: Intellipaat, "Tableau Architecture," *https://intellipaat.com/blog/tutorial/tableau-tutorial/tableau-architecture/*

Other reporting tools to explore include Qlik, Oracle, SAP and Alteryx.

Once reporting is completed, it is critical to continuously update the models and tweak data processing to further refine the models and optimize the resulting recommended actions. Whether personal or business, goals are either achieved or further modified to stay abreast of changes in the market and remain competitive. Unless updated and consistently managed, any data asset can become stale and lose its value proposition to the organization.

3.7 Chapter 3 Knowledge Check

REVIEW QUESTIONS

1. Choose the **BEST** answer. Prior to starting data processing, the following should be done first:

   A. Gathering input data samples
   B. Preprocessing data
   C. Creating clear, concise requirements
   D. Formulating a good hypothesis

2. When data mining the web, the following method is **MOST** often used by search engines:

   A. Scraping
   B. Machine learning
   C. Crawling
   D. Artificial intelligence algorithms

3. Partitioning, hierarchical, fuzzy and dense-based are all examples of:

   A. Supervised classification
   B. Data classification
   C. Data clustering methods
   D. Machine learning algorithms

4. Predictive analytics asks:

   A. Why did that happen?
   B. When did it happen?
   C. What should we do?
   D. What could happen?

5. Choose the **BEST** answer. When communicating results to stakeholders, it is best:

   A. To use a Boston Consulting Group (BCG) graph
   B. To practice your presentation
   C. To know your audience and tailor your presentation
   D. To incorporate a kinesthetic approach

Answers on page 130

Chapter 3 ANSWER KEY

Review Questions

1. A. Gathering input data samples
   B. Preprocessing data
   C. Creating clear, concise requirements
   D. Formulating a good hypothesis. Refer to pages 93-94.

2. A. Scraping.
   B. Machine learning
   C. Crawling. Refer to page 96.
   D. Artificial intelligence algorithms

3. A. Supervised classification
   B. Data classification
   C. Data clustering methods. Refer to pages 99-101.
   D. Machine learning algorithms

4. A. Why did that happen?
   B. When did it happen?
   C. What should we do?
   D. What could happen? Refer to page 112.

5. A. To use a Boston Consulting Group (BCG) graph
   B. To practice your presentation
   C. To know your audience and tailor your presentation. Refer to page 120.
   D. To incorporate a kinesthetic approach

Glossary

A

A/B Testing—A statistical way of comparing two (or more) techniques, typically an incumbent against a new rival. A/B testing aims to determine not only which technique performs better but also whether the difference is statistically significant. A/B testing usually considers only two techniques using one measurement but can be applied to any finite number of techniques and measures.

Accuracy—The fraction of predictions that a classification model got right. In multi-class classification, accuracy is defined as correct predictions divided by total number of examples. In binary classification accuracy is defined as (true positives + true negatives) divided by total number of examples.

Action—In reinforcement learning, the mechanism by which the agent transitions between states of the environment. The agent chooses the action by using a policy.

Active learning—A training approach in which the algorithm chooses some of the data it learns from. Active learning is particularly valuable when labeled examples are scarce or expensive to obtain. Instead of blindly seeking a diverse range of labeled examples, an active learning algorithm selectively seeks the particular range of examples it needs for learning.

AdaGrad—A sophisticated gradient descent algorithm that rescales the gradients of each parameter, effectively giving each parameter an independent learning rate

Algorithm—A finite set of well-defined, unambiguous rules for the solution of a problem in a finite number of steps, it is a sequence of operational actions that lead to a desired goal and is the basic building block of a program

AngularJS—An open-source JavaScript library maintained by Google and the AngularJS community that lets developers create what are known as Single [web] Page Applications. AngularJS is popular with data scientists as a way to show the results of their analysis.

Artificial intelligence—An advanced computer system that can simulate human capabilities, such as analysis, based on a predetermined set of rules

Average precision—A metric for summarizing the performance of a ranked sequence of results. Average precision is calculated by taking the average of the precision values for each relevant result (each result in the ranked list where the recall increases relative to the previous result).

B

Backpropagation—An algorithm for iteratively adjusting the weights used in a neural network system. Backpropagation is often used to implement gradient descent.

Bayes' Theorem—An equation for calculating the probability that something is true if something potentially related to it is true. If P(A) means "the probability that A is true" and P(A|B) means "the probability that A is true if B is true," then Bayes' Theorem tells us that $P(A|B) = (P(B|A)P(A)) / P(B)$.

Bayesian network—Graphs that compactly represent the relationship between random variables for a given problem. These graphs aid in performing reasoning or decision making in the face of uncertainty. These networks are usually represented as graphs in which the link between any two nodes is assigned a value representing the probabilistic relationship between those nodes.

Bias—In machine learning, bias is a learner's tendency to consistently learn the same wrong thing.

Scope Notes: Variance is the tendency to learn random things irrespective of the real signal. For example, it is easy to avoid overfitting (variance) by falling into the opposite error of underfitting (bias).

Big data—The ability to work with collections of data that had been impractical before because of their volume, velocity, and variety ("the three Vs"). A key driver of this new ability has been easier distribution of storage and processing across networks of inexpensive commodity hardware using technology such as Hadoop instead of requiring larger, more powerful individual computers.

Binomial distribution—A distribution of outcomes of independent events with two mutually exclusive possible outcomes, a fixed number of trials, and a constant probability of success. This is a discrete probability distribution, as opposed to continuous.

Boosting—A machine learning technique that iteratively combines a set of simple and not very accurate classifiers (referred to as "weak" classifiers) into a classifier with high accuracy (a "strong" classifier) by upweighting the examples that the model is currently mis-classifying

C

Calibration layer—A post-prediction adjustment, typically to account for prediction bias. The adjusted predictions and probabilities should match the distribution of an observed set of labels.

Candidate generation—The initial set of recommendations chosen by a recommendation system

Centroid—The center of a cluster as determined by a k-means or k-median algorithm. For instance, if k is 3, then the k-means or k-median algorithm finds 3 centroids.

Checkpoint—Data that capture the state of the variables of a model at a particular time. Checkpoints enable exporting model weights and performing training across multiple sessions. Checkpoints also enable training to continue past errors (e.g., job preemption).

Chi-square test—An analysis technique used to estimate whether two variables in a cross tabulation are correlated. A chi-square distribution varies from normal distribution based on the "degrees of freedom" used to calculate it.

Classification—The identification of two or more categories in which an item belongs; a classic machine learning task

Clipping—A technique for handling outliers. Specifically, reducing feature values that are greater than a set maximum value down to that maximum value. Also, increasing feature values that are less than a specific minimum value up to that minimum value.

Clustering—An algorithm for dividing data instances into groups—not a predetermined set of groups, which would make this classification, but groups identified by the execution of the algorithm because of similarities that it found among the instances. The center of each cluster is known as "centroid."

Co-adaptation—When neurons predict patterns in training data by relying almost exclusively on outputs of specific other neurons instead of relying on the network's behavior as a whole

Coefficient—A number or algebraic symbol prefixed as a multiplier to a variable or unknown quantity (Ex.: x in x(y + z), 6 in 6ab)

Collaborative filtering—Making predictions about the interests of one user based on the interests of many other users. Collaborative filtering is often used in recommendation systems.

Computational linguistics—A branch of computer science for parsing text of spoken languages (e.g., English or Mandarin) to convert it to structured data that can be used to drive program logic

Confidence interval—A range specified around an estimate to indicate margin of error, combined with a probability that a value will fall in that range

Continuous feature—A floating-point feature with an infinite range of possible values. Contrasts with discrete feature.

Continuous variable—A variable whose value can be any of an infinite number of values, typically within a particular range

Convenience sampling—Using a dataset not gathered scientifically in order to run quick experiments. Later on, it is essential to switch to a scientifically gathered dataset.

Convergence—Informally, often refers to a state reached during training in which training loss and validation loss change very little or not at all with each iteration after a certain number of iterations

Correlation—The degree of relative correspondence between two sets of data. The correlation coefficient is a measure of how closely the two data sets correlate.

Covariant—A measure of the relationship between two variables whose values are observed at the same time. Whereas variance measures how a single variable deviates from its mean, covariance measures how two variables vary in tandem from their means.

Crash blossom—A sentence or phrase with an ambiguous meaning

Cross-validation—A mechanism for estimating how well a model will generalize to new data by testing the model against one or more nonoverlapping data subsets withheld from the training set

D

D3—Data-Driven Documents. A JavaScript library that eases the creation of interactive visualizations embedded in web pages. D3 is popular with data scientists as a way to present the results of their analysis.

Data—Representations of facts, concepts or instructions in a manner suitable for communication, interpretation or processing by humans or by automated means. In the simplest terms, data are pieces of information.

Data analysis—Obtaining an understanding of data by considering samples, measurement and visualization. Data analysis can be particularly useful when a data set is first received, before one builds the first model. It is also crucial in understanding experiments and debugging problems with the system.

Data augmentation—Artificially boosting the range and number of training examples by transforming existing examples to create additional examples

Data dictionary—Stores all of the details that corresponds to the DFD stores, processes and flows. It may be called a database that contains the name, type, range of values, source and authorization for access for each data element in a system. It also indicates which application programs use those data so that when a data structure is contemplated, a list of the affected programs can be generated.

Data mining—Generally, the use of computers to analyze large data sets to look for patterns that assist people in making business decisions

Data science—A new branch of science used to extract knowledge and insights from large and complex data sets. Data science work often requires knowledge of both statistics and software engineering.

Data structure—A particular arrangement of units of data such as an array or a tree

Data wrangling—The conversion of data, often through the use of scripting languages, to make it easier to work with

DataFrame—A popular datatype for representing datasets in pandas. A DataFrame is analogous to a table. Each column of the DataFrame has a name (a header), and each row is identified by a number.

Decision boundary—The separator between classes learned by a model in a binary class or multi-class classification problems

Decision trees—A tree structure to represent a number of possible decision paths and an outcome for each path

Deep learning—A multi-level algorithm that gradually identifies things at higher levels of abstraction. For example: image classification.

Deep model—A type of neural network containing multiple hidden layers

Dependent variable—The value of a dependent value "depends" on the value of the independent variable

Dimension reduction—A technique to extract one or more dimensions that capture as much of the variation in the data as possible

Dimensionality—In statistics it refers to how many attributes a dataset has

Discrete variable—A variable whose potential values must be one of a specific number of values. Also known as discrete feature.

Discriminative model—A model that predicts labels from a set of one or more features. More formally, discriminative models define the conditional probability of an output given the features and weights.

Discriminator—A system that determines whether examples are real or fake

Downsampling—Reducing the amount of information in a feature in order to train a model more efficiently

Dynamic model—A model that is trained online in a continuously updating fashion. That is, data is continuously entering the model.

E

Early stopping—A method for regularization that involves ending model training before training loss finishes decreasing

Econometrics—The use of mathematical and statistical methods in the field of economics to verify and develop economic theories.

Episode—In reinforcement learning, each of the repeated attempts by the agent to learn an environment

Epoch—A full training pass over the entire dataset such that each example has been seen once. Thus, an epoch represents N/batch size training iterations, where N is the total number of examples.

Epsilon greedy policy—In reinforcement learning, a policy that either follows a random policy with epsilon probability or a greedy policy otherwise

Exploding gradient problem—The tendency for gradients in a deep neural networks (especially recurrent neural networks) to become surprisingly steep (high)

F

False negative—In intrusion detection, an error that occurs when an attack is misdiagnosed as a normal activity

False positive (FP)—An example in which the model mistakenly predicted the positive result

Feature—The machine-learning expression for a piece of measurable information about something. If you store the age, annual income and weight of a set of people, you are storing three features about them.

Feature cross—A synthetic feature formed by crossing (taking a Cartesian product of) individual binary features obtained from categorical data or from continuous features via bucketing. Feature crosses help represent nonlinear relationships.

Federated learning—A distributed machine-learning approach that trains machine-learning models using decentralized examples residing on devices, such as smartphones

Feedforward neural network (FFN)—A neural network without cyclic or recursive connections. For example, traditional deep neural networks are feedforward neural networks.

Few-shot learning—A machine-learning approach, often used for object classification, designed to learn effective classifiers from only a small number of training examples

G

Garbage in, garbage out (GIGO)—The concept of data that is nonsensical, or flawed, especially as it relates to the computational sciences

General Architecture for Text Engineering (GATE)—An open source, Java-based framework for natural language processing tasks. The framework lets developers pipeline other tools designed to be plugged into it. The project is based at the UK University of Sheffield.

Generalization—Refers to the ability of the model to make correct predictions on new, previously unseen data, as opposed to the data used to train the model

Gradient boosting—A machine-learning technique for regression and classification problems that produces a prediction model in the form of an ensemble of weak prediction models, typically decision trees. It builds the model in a stage-wise fashion, like other boosting methods, and it generalizes them by allowing optimization of an arbitrary differentiable loss function.

Gradient descent—An optimization algorithm for finding the input to a function that produces the largest (or smallest) possible value

Greedy policy—In reinforcement learning, a policy that always chooses the action with the highest expected return

Ground truth—The correct answer; reality. Since reality is often subjective, expert raters typically are the proxy for ground truth.

H

Hashing—In machine learning, a mechanism for bucketing categorical data, particularly when the number of categories is large, but the number of categories actually appearing in the dataset is comparatively small.

Heuristic—A quick solution to a problem, which may or may not be the best solution

Hidden layer—A synthetic layer in a neural network between the input layer (i.e., the features) and the output layer (the prediction). Hidden layers typically contain an activation function (e.g., ReLU) for training. A deep neural network contains more than one hidden layer.

Histogram—A graphical representation of the distribution of a set of numeric data, usually a vertical bar graph

Hyperparameter—The knobs that are tweaked during successive runs of training a model

I

Image recognition—A process that classifies objects, patterns or concepts in an image. Image recognition is also known as image classification.

Independently and identically distributed (i.i.d)—Data drawn from a distribution that does not change, and where each value drawn does not depend on values that have been drawn previously. An i.i.d. is the ideal gas of machine learning—a useful mathematical construct but almost never exactly found in the real world.

Inference—In machine learning, often refers to the process of making predictions by applying the trained model to unlabeled examples

J

JavaScript—A scripting language originally designed in the mid-1990s for embedding logic in web pages, but which later evolved into a more general-purpose development language. JavaScript continues to be very popular for embedding logic in web pages.

K

K-means clustering—A data-mining algorithm to cluster, classify or group N objects based on their attributes or features into K number of groups (so-called clusters)

K-nearest neighbors—A machine-learning algorithm that classifies things based on their similarity to nearby neighbors. The algorithm execution is refined by picking how many neighbors to examine (k) and some notion of distance to indicate how near the neighbors are.

Keras—A popular Python machine learning API

Key performance indicator (KPI)—A measure that determines how well the process is performing in enabling the goal to be reached

Scope Notes: A lead indicator of whether a goal will likely be reached, and a good indicator of capabilities, practices and skills. It measures an activity goal, which is an action that the process owner must take to achieve effective process performance.

Keypoints—The coordinates of particular features in an image

L

Label—In supervised learning, the answer or result portion of an example

Latent variable—Variables that are not directly observed, but are rather inferred (through a mathematical model) from other variables that are observed (directly measured)

Lift—Compares the frequency of an observed pattern with how often one expects to see that pattern just by chance

Linear algebra—A branch of mathematics dealing with vector spaces and operations on them, such as addition and multiplication. It is designed to represent systems of linear equations.

Linear regression—A technique to look for a linear relationship (i.e., one where the relationship between two varying amounts, such as price and sales, can be expressed with an equation that can be represented as a straight line on a graph) by starting with a set of data points that do not necessarily line up nicely

Logistic regression—A model similar to linear regression but where the potential results are a specific set of categories, instead of being continuous

M

Machine learning—A program or system that builds (i.e., trains) a predictive model from input data. The system uses the learned model to make useful predictions from new (never-before-seen) data drawn from the same distribution as the one used to train the model. Machine learning also refers to the field of study concerned with these programs or systems.

Machine learning model—The process of training a machine learning model involves providing a machine learning model algorithm (that is, the learning algorithm) with training data to learn from. The term machine learning model refers to the model artifact that is created by the training process.

Markov Chain—An algorithm for working with a series of events (for example, a system being in particular states) to predict the possibility of a certain event based on which other events have happened. The identification of probabilistic relationships between the different events.

Markov decision process (MDP)—A graph representing the decision-making model where decisions (or actions) are taken to navigate a sequence of states, under the assumption that the Markov property holds. In reinforcement learning, these transitions between states return a numerical reward.

MATLAB—A commercial computer language and environment popular for visualization and algorithm development

Matrix—A set of numbers or terms arranged in rows and columns between parentheses or double lines. For purposes of manipulating a matrix with software, think of it as a two-dimensional array. As with its one-dimensional equivalent, a vector, this mathematical representation of the two-dimensional array makes it easier to take advantage of software libraries that apply advanced mathematical operations to the data—including libraries that can distribute the processing across multiple processors for scalability.

Mean—The average value, also known as arithmetic mean

Mean absolute error—The average error of all predicted values when compared with observed values

Mean squared error—The average of the squares of all the errors found when comparing predicted values with observed values

Median—When values are sorted, the value in the middle, or the average of the two in the middle if there are an even number of values

Mode—The value that occurs most often in a sample of data. Like the median, the mode cannot be directly calculated.

Model—A way to describe a given set of components and how those components relate to each other to describe the main workings of an object, system or concept

Monte Carlo method—The use of randomly generated numbers as part of an algorithm

Moving average—The mean (or average) of time series data (observations equally spaced in time, such as per hour or per day) from several consecutive periods is called the moving average

N

N-gram—The analysis of sequences of n items (typically, words in natural language) to look for patterns. The value of n can be anything. This is used to construct statistical models of documents (for example, when automatically classifying them) and to find positive or negative terms associated with a product name.

Naive Bayes classifier—A collection of classification algorithms based on Bayes Theorem. It is a family of algorithms that all share a common principle that every feature being classified is independent of the value of any other feature.

NaN trap—When one number in your model becomes a NaN during training, which causes many or all other numbers in your model to eventually become a NaN. NaN is an abbreviation for Not a Number.

Neural network—A robust function that takes an arbitrary set of inputs and fits it to an arbitrary set of outputs that are binary. Neural networks are used in deep learning research to match images to features and much more.

Neuron—A node in a neural network, typically taking in multiple input values and generating one output value

Node (neural network)—A neuron in a hidden layer

Noise—Disturbances in data transmissions or data set, such as static, that cause messages to be misinterpreted by the receiver

Normal distribution—A probability distribution that, when graphed, is a symmetrical bell curve with the mean value at the center. The standard deviation value affects the height and width of the graph. Also known as Gaussian distribution.

Normalization—The process of converting an actual range of values into a standard range of values, typically -1 to +1 or 0 to 1.

NoSQL—A database management system that uses any of several alternatives to the relational, table-oriented model used by SQL databases

Null hypothesis—If the proposed model for a data set says that the value of x is affecting the value of y, then the null hypothesis—the model with which you are comparing your proposed model to check whether x really is affecting y—says that the observations are all based on chance and that there is no effect. The smaller the P-value computed from the sample data, the stronger the evidence is against the null hypothesis.

O

Objective—Statement of a desired outcome

Objective function—Used to solve an optimization problem by combining decision variables, constraints and the goal value into an objective function. The objective is the goal desired to maximize or minimize; the objective function is used to find the optimum result.

Observation—Receipt of messages through electronic, sensory or vibrational signals, and the human senses

Offline inference—Generating a group of predictions, storing those predictions and then retrieving those predictions on demand

One-shot learning—A machine-learning approach, often used for object classification, designed to learn effective classifiers from a single training example

Outlier—Extreme values that might be errors in measurement and recording, or might be accurate reports of rare events

Overfitting—A model of training data that, by taking too many of the data quirks and outliers into account, is overly complicated and will not be as useful as it could be to find patterns in test data

P

P value—The probability, under the assumption of no effect or no difference (the null hypothesis), of obtaining a result equal to or more extreme than what was actually observed

PageRank—An algorithm that determines the importance of something, typically to rank it in a list of search results. PageRank works by counting the number and quality of links to a page to determine a rough estimate of how the importance of the website. The underlying assumption is that more important websites are likely to receive more links from other websites.

Pandas—A Python library for data manipulation that is popular with data scientists

Perceptron—Neural network that approximates a single neuron with n binary inputs. It computes a weighted sum of its inputs and fires if that weighted sum is zero or greater.

Perplexity—One measure of how well a model is accomplishing its task

Pivot table—Pivot tables quickly summarize long lists of data, without requiring a single formula or copy a single cell. But the most notable feature of pivot tables is that they can be arranged dynamically.

Poisson distribution—A distribution of independent events, usually over a period of time or space, used to help predict the probability of an event. Like the binomial distribution, this is a discrete distribution.

Predictive analytics—The analysis of data to predict future events, typically to aid in business planning. Predictive analytics incorporates predictive modeling and other techniques. Machine learning may be considered a set of algorithms to help implement predictive analytics.

Predictive modeling—The development of statistical models to predict future events

Preprocessing—Processing data before it is used to train a model

Principal component analysis—This algorithm simply looks at the direction with the most variance and then determines that as the first principal component. This is very similar to how regression works in that it determines the best direction to which to map data.

Prior distribution—In Bayesian inference, it is assumed that the unknown quantity to be estimated has many plausible values modeled by a prior distribution. Bayesian inference is then using data (that is considered as unchanging) to build a tighter posterior distribution for the unknown quantity.

Probability distribution—A probability distribution for a discrete random variable is a listing of all possible distinct outcomes and their probabilities of occurring. Because all possible outcomes are listed, the sum of the probabilities must add to 1.0.

Proxy (sensitive attributes)—An attribute used as a stand-in for a sensitive attribute

Python—A scripting programming language available since 1994 that is popular for data science

Q

Quantile, quartile—When a set of sorted values is divided into groups that each have the same number of values (for example, if the values are divided into two groups at the median), each group is known as a quantile. If values are divided into four groups, they are called quartiles, which is a common way to divide values for discussion and analysis purposes. If there are five groups, they are called quintiles, and so forth.

R

R—An open-source programming language and environment for statistical computing and graph generation available for Linux, Windows and Mac

Random forest—An ensemble approach to finding the decision tree that best fits the training data by creating many decision trees and then determining the average one. The random part of the term refers to building each of the decision trees from a random selection of features; the forest refers to the set of decision trees.

Rank (ordinality)—The ordinal position of a class in a machine-learning problem that categorizes classes from highest to lowest

Rater—A human who provides labels in examples. Sometimes called an annotator.

Recall—A metric for classification models that answers the following question: Out of all the possible positive labels, how many did the model correctly identify?

Recurrent neural network—A neural network that is intentionally run multiple times, where parts of each run feed into the next run

Regression model—A type of model that outputs continuous (typically, floating-point) values

Reinforcement learning—A class of machine-learning algorithms in which the process is not given specific goals to meet but, as it makes decisions, is instead given indications of whether it is doing well or not

Root Mean Squared Error—The square root of the Mean Squared Error

Ruby—A scripting language that first appeared in 1996. Ruby is popular in the data science community, but not as popular as Python, which has more specialized libraries available for data science tasks.

S

S curve—A type of curve that shows the growth of a variable in terms of another variable, often expressed as units of time. The S curve is often mentioned when someone predicts that a rising value will eventually level off.

SAS—A commercial statistical software suite that includes a programming language also known as SAS

Scalar—A quantity that has magnitude but no direction in space, as volume or temperature

Scaling—A commonly used practice in feature engineering to tame the range of values of a feature to match the range of other features in the data set

Scoring—The part of a recommendation system that provides a value or ranking for each item produced by the candidate generation phase

Scripting—Generally, the use of a computer language to write a program or script that can be run directly, with no need to compile it to binary code, as with languages such as Python, Java and C

Sensitive attribute—A human attribute that may be given special consideration for legal, ethical, social or personal reasons

Serial correlation—The relationship between a variable and a lagged version of itself over various time intervals. Repeating patterns often show serial correlation when the level of a variable affects its future level.

Shell—Command line scripting languages, such as Perl and Python. Linux-based shell tools (which are either included with or easily available for Mac and Windows machines), such as grep, diff, split, comm, head and tail, are popular for data wrangling

Spatiotemporal data—Time series data that also includes geographic identifiers, such as latitude-longitude pairs

SPSS—A commercial statistical software package used for predictive analysis

SQL—The ISO standard query language used by application programmers and end users to access relational databases. Variations of this popular language are often available for data storage systems that are not strictly relational.

Standard deviation—The square root of the variance, and a common way to indicate just how different a particular measurement is from the mean

Standard normal distribution—A normal distribution with a mean of 0 and a standard deviation of 1. When graphed, it is a bell-shaped curve that is centered around the y axis, where x=0.

Standardized score—Transforms a raw score into units of standard deviation above or below the mean. This translates the scores so they can be evaluated in reference to the standard normal distribution.

Stata—A commercial statistical software package, not to be confused with strata

Strata, stratified sampling—Sampling technique used to divide the units into homogeneous groups (strata) and draw a simple random sample from each group

Supervised learning—A type of machine learning algorithm in which a system is taught to classify input into specific, known classes

T

T-distribution—A variation on normal distribution that accounts for the fact that only a sampling of all the possible values is being used instead of all of them

Tableau—A commercial data visualization package often used in data science projects

TensorFlow—A large-scale, distributed, machine-learning platform

Test set—The subset of the data set used to test a model after the model has gone through initial vetting by the validation set

Time series data—Time series data have measurements of observations accompanied by date-time stamps

Training—The process of determining the ideal parameters comprising a model

U

UIMA—The Unstructured Information Management Architecture framework is used to analyze unstructured information, especially natural language. OASIS UIMA is a specification that standardizes this framework, and Apache UIMA is an open-source implementation of it.

Unsupervised learning—A class of machine-learning algorithms designed to identify groupings of data without knowing in advance what the groups will be

V

Variance—How much a list of numbers varies from the mean (average) value. It is frequently used in statistics to measure how large the differences are in a set of numbers. It is calculated by averaging the squared difference of every number from the mean.

Vector—An ordered set of real numbers, each denoting a distance on a coordinate axis. These numbers may represent a series of details about a single person, movie, product or whatever entity is being modeled.

Vector space—A collection of vectors, e.g., a matrix

W

Weight—A coefficient for a feature in a linear model, or an edge in a deep network

Weka—An open-source set of command line and graphical user interface data analysis tools developed at the University of Waikato, in New Zealand

Width—The number of neurons in a particular layer of a neural network

Page intentionally left blank

9781604207552